Recent Research in Psychology

Leslie P. Steffe
Editor

Epistemological Foundations of Mathematical Experience

With 50 Illustrations

Springer-Verlag
New York Berlin Heidelberg London
Paris Tokyo Hong Kong Barcelona

1991

Leslie P. Steffe
The University of Georgia
College of Education
Mathematics Education Department
Athens, GA 30602
USA

Library of Congress Cataloging-in-Publication Data
Epistemological foundations of mathematical experience / edited by
Leslie P. Steffe
p. cm.
Includes bibliographical references and index.
ISBN 0-387-97600-0
1. Mathematics—Study and teaching—Psychological aspects—
Congresses. I. Steffe, Leslie P.
QA11.A1E64 1991
510'.7'1—dc20 91-15438

Printed on acid-free paper.

Camera-ready copy provided by the editor using Microsoft Word.
Printed and bound by Edwards Brothers, Inc., Ann Arbor, MI.
Printed in the United States of America.

9 8 7 6 5 4 3 2 1

ISBN 0-387-97600-0 Springer-Verlag New York Berlin Heidelberg
ISBN 3-540-97600-0 Springer-Verlag Berlin Heidelberg New York

Dedicated to the Memory of Myron F. Rosskopf

Preface

On the 26th, 27th, and 28th of February of 1988, a conference was held on the epistemological foundations of mathematical experience as part of the activities of NSF Grant No. MDR-8550463, *Child Generated Multiplying and Dividing Algorithms: A Teaching Experiment*. I had just completed work on the book *Construction of Arithmetical Meanings and Strategies* with Paul Cobb and Ernst von Glasersfeld and felt that substantial progress had been made in understanding the early numerical experiences of the six children who were the subjects of study in that book. While the book was in preparation, I was also engaged in the teaching experiment on multiplying and dividing algorithms. My focus in this teaching experiment was on investigating the mathematical experiences of the involved children and on developing a language through which those experiences might be expressed. However, prior to immersing myself in the conceptual analysis of the mathematical experiences of the children, I felt that it was crucial to critically evaluate the progress that we felt we had made in our earlier work. It was toward achieving this goal that I organized the conference.

When trying to understand the mathematical experiences of a child, one can do no better than to interact with the child in a mathematical context guided by the intention to specify the child's current knowledge and the progress the child might make. Mathematical experiences, being relative to the particular interaction in which they occur, are always in flux and are as much a function of the adult's as they are of the child's intentions, language, and actions. Interactive mathematical activity is certainly not a neutral datum. The adult's understanding of children's interactive mathematical experiences form the foundation of any language through which children's experiences are expressed. Hence, if we, the observers, hope to communicate with each other a consensual but dynamically changing and expanding domain of interpretive constructs needs to be established. It is critical to know how the constructs that we already take as shared are being modified through their use as well as what other constructs are being developed that are yet to be shared.

Piaget's reflective abstraction was one of the taken-as-shared interpretive constructs among the conferees. I was interested in how the other investigators at the conference thought about and used reflective abstraction because I have found it possible to observe children in the process of making a reflective abstraction in an experiential context. Did they view it in a functional way or did they restrict it to only the most general reorganizations that accompany discontinuities in mathematical learning? I invited Ernst von Glasersfeld to finish a paper on reflective abstraction that he had begun earlier in which he interpreted reflective abstraction in the context of scheme theory. His theoretical analysis had opened up the possibility for me to view reflective abstraction as a relevant "everyday concept" and my hope was that von Glasersfeld's paper would encourage others to view it in that way.

In order to explore how reflective abstraction is used by investigators in fields other than mathematics, I invited Philip Lewin to discuss his work in education in the humanities and social sciences. In his chapter, he stresses both reflective abstraction and the role of guiding images in the organization of knowing in domains that he believes are of necessity fuzzy, and where the prior experience and belief systems of the knower--even the expert--frequently have more salience than in the hard sciences and mathematics. During his presentation at the conference, he involved all of us in making reflective abstractions in the humanities and we came away with an experience that will not be soon forgotten.

Ed Dubinsky shows us in his chapter just how important the prior experiences and belief systems of the knower are in advanced mathematical thinking as he discusses what he takes to be salient aspects of reflective abstraction. Along with Lewin, he provides contexts where terms like "encapsulation" and "interiorization" take on functional meaning. At the other end of the spectrum of mathematical thinking, I discuss these processes in terms of a feedback system, where the children's reflective abstractions use current sensory-motor and figurative experiences as material. The belief that mathematics is a product of reflective abstraction is developed in these four papers. No longer is reflective abstraction a theoretical construct that is used to explain only the most general and unobservable changes in the nature and organization of mathematical knowledge. Rather, it is an indispensable tool in explaining how children modify their mathematical experiences.

Recasting reflective abstraction into functional forms is crucial if constructivism is to realize its potential in revitalizing the field of mathematics education. There must be a theory of construction that explains how mathematical operations are introduced by the actor as novelties through the actor's interactions with elements in its environment and how those novelties are lifted from their experiential forms. Mark Bickhard shows us how interactivism leads to a constructivism and how it accounts for reflective abstraction. This is another way of saying that reflective abstraction cannot be taken as a given in any organism nor can any of its products be taken as a given, including mathematics. Rather, reflective abstraction is itself introduced by the actor as a novelty and consists of combinations of mental operations. How these operations, once introduced, become available to the actor in any particular context is a critical research problem whose resolution would be of great benefit to education in mathematics. My current hypothesis is that reflective abstraction is not ubiquitous nor is it independent of current mathematical knowledge or of particular environments. In view of Bickhard's work on rationality, we should expect our models of reflective abstraction to contain what he calls principles of selection (critical principles), and because of the similarity of critical principles and negative feedback in scheme theory, the work of von Glasersfeld and that of Bickhard are reciprocally informing.

Interactivism finds an expression in the paper by Robert G. Cooper, where the idea of repeated experience is explored (repeated interaction of an organism with its environment). When I invited him to the conference I had no idea how solidly his thinking would connect with a functional view of reflective abstraction and with the work on recursion discussed by Thomas Kieren and Susan E.B. Pirie. As Kieren and Pirie explain it, recursion is based on an interactivism. So, along with feedback in the context of scheme theory, it can be used to broaden our understanding of repeated experience as advanced by Cooper.

Repeated interaction of an organism with its environment, when applied to us as researchers, is a crucial part of constructing a conceptual map of children's mathematical experiences. I have found learning children's mathematics to be a very protracted process with the child as my only teacher. Although I am always interested in the modifications that children make in their mathematical knowledge as they work with me, without a model of that knowledge the modifications could be only described, not explained. Jere Confrey, in her paper, captures the essence of the models in a comment that I find particularly elegant in its simplicity--"Through the process of the interview, my own conception of exponential functions was transformed, elucidated, and enriched" (p. 129). This comment highlights the relativistic nature of our understanding of the mathematics of another person--we understand that mathematics using our own conceptual constructs. So if we take learning the mathematical knowledge of another person seriously, we should expect modifications in our knowledge to occur. Confrey uses the operation of unitizing--encapsulation in the language of Dubinsky and Lewin--throughout her paper to explain a student's concept of multiplication. We also find recursion as explained by Kieren and Pirie re-emerging as a way to interpret the student's work with exponential growth or decay. These examples serve as confirmations of the functional nature of reflective abstraction and contribute to the emergence of a new learning theory in mathematics education.

Larry Hatfield introduces the idea of a high quality mathematical experience. He identifies emotional states and enquiry states--states of feeling and thinking--as inter-related but distinguishable in his explorations of what it means to have high quality mathematical experiences. These two states of feeling and thinking are what Bickhard calls critical principles. Bickhard believes, as does Hatfield, that rationality emerges out of creative engagement with problems. They both develop the thesis that the standard opposition between the emotions and the thought of an individual is a false opposition. Without motivation, interactivism ceases to be a viable theory of knowing and the individual necessarily is viewed as a passive organism waiting for the world to impress itself on his or her mind. Such a regressive view of mind is discussed in the introductory chapter written by Clifford Konold and David K. Johnson.

The introductory and the final discussion chapter written by Patrick W. Thompson were not available at the time of the conference in any form. Konold, Thompson, and von Glasersfeld served as discussants of the other papers in this volume and it was only after the conference that they graciously accepted my invitation to write their respective chapters. As editor of the volume, I wish to express my appreciation to them and to David K. Johnson for coauthoring the introductory chapter.

Several people served as discussion moderators and contributed immensely to the quality of the conference through their commentaries and syntheses of the discussions. As conference director, I express my gratitude to Dr.'s Ben Blount, Paul Cobb, Pat Kyllonen, John Olive, Neil Pateman, George Stanic, and Patricia Wilson. Finally, I would like to thank the staff of the Institute for Behavioral Research and the Department of Mathematics Education for their help in organizing and conducting the conference and in preparing the manuscript for this book. Special thanks are due to Ms. Gilda A. Ivory for her cheerful and tireless help in typing the many revisions of the papers and in preparing a camera-ready copy of the manuscript.

Leslie P. Steffe
Athens, Georgia

Acknowledgments

The conference on which this book is based was supported by the National Science Foundation under Grant No. MDR-85504063, and the Department of Mathematics Education and the Institute for Behavioral Research both of the University of Georgia. All opinions and findings are those of the authors' and are not necessarily representative of the sponsoring agencies.

Contents

Contributors

Mark H. Bickhard
Department of Psychology
Lehigh University
Bethlehem, PA 18015, USA

Jere Confrey
Department of Education
Cornell University
Ithaca, NY 14853, USA

Robert G. Cooper, Jr.
Department of Psychology
San Jose State University
San Jose, CA 95125, USA

Ed Dubinsky
Department of Education
Purdue University
West LaFayette, IN 47907, USA

Ernst von Glasersfeld
37 Long Plain Rd.
RFD 3
Amherst, MA 01002, USA

Larry L. Hatfield
Mathematics Education
University of Georgia
Athens, GA 30602, USA

David K. Johnson
Academic Foundations
School of Education
Rutgers University
Newark, NJ 07102, USA

Thomas E. Kieren
5208 142nd St.
Edmonton, Alberta T6H 4B4, Canada

Clifford Konold
256 North Silver Lane
Sunderland, MA 01375, USA

Philip Lewin
Liberal Studies
Carkson University
Potsdam, NY 13676, USA

Susan E.B. Pirie
Director
Mathematics Education Research
Center
Department of Educational Studies
Oxford University
Oxford, OX2 6PY
England

Patrick W. Thompson
Department of Mathematics
CRMEE
SDSU
San Diego, CA 92182, USA

1 Philosophical and Psychological Aspects of Constructivism

Clifford Konold and David K. Johnson

> For whatever may be said about the importance of aiming at depth rather than width in our studies, and however the demand of the present age may be for specialists, there will always be work, not only for those who build up particular sciences and write monographs on them, but for those who open up such communications between the different groups of builders as will facilitate a healthy interaction between them.
>
> James Clerk Maxwell (1878)

Clifford Geertz (1983) speaks of "genre blurring" to refer to, among other things, the cross-fertilization of the social sciences and the humanities. In the process, the social sciences are giving up their long-held objective of patterning themselves after the physical sciences. This volume might be looked at as a case study of such genre blurring. Some of the contributors come from academic backgrounds other than mathematics and mathematics education. They include academics trained in psychology, philosophy, and classical studies. This diversity is reflected to some extent in the lack of overlap in the works each chapter references. But there is a stronger rationale for characterizing these chapters as cross-disciplinary: one finds within single chapters, references culled from a variety of disciplines. Although the authors come from and draw on diverse disciplines, they share a core perspective. This shared perspective is a species of constructivism.

There are many possible tacks to take in an introductory chapter such as this. What seems the most appropriate in this case is to orient the reader who is interested in mathematics education, for example, but who may be unaware of constructivist epistemology. Accordingly, we describe general philosophical and psychological issues that underlie the constructivism advocated by these authors. Having done this, we provide a brief overview of the volume in which we highlight the primary focus of each chapter.

Foundationalism and Constructivism

> It is now some years since I detected how many were the false beliefs that I had from my earliest youth admitted as true, and how doubtful was everything I had since constructed on this basis; and from that time I was convinced that I must once and for all seriously undertake to rid myself of all the opinions which I had formerly accepted, and commence to build anew from the foundation, if I wanted to establish any firm and permanent structure in the sciences.
>
> Descartes, Meditations I

One way to capture the temper of the modern age is to invoke Richard Bernstein's (1983) phrase "Cartesian anxiety." It is an anxiety that permeates all metaphysical and epistemological questions concerning the existence of a stable and reliable rock upon which we can secure our thoughts and actions. As Bernstein explains: "Either there is some support for our being, a fixed foundation for our knowledge, or we cannot escape the forces of darkness that envelop us with madness, with intellectual and moral chaos" (p. 18).

The work of Descartes marked the beginning of what philosophers now refer to as the "epistemological turn" (away from metaphysics, or questions about being). After Descartes, the essential problem facing philosophers was to account for the possibility of knowledge within the context of a radical subject-object dichotomy. This duality followed from the division of reality into two nonintersecting realms: an "inner" realm of the subject's knowledge and an "outer" or external realm consisting of the objects of that knowledge. The problem of identifying the possibility of (true) knowledge, given the additional Cartesian assumption that knowledge must somehow conform to the object (and not vice versa), becomes the problem of accounting for the possibility that the subject can "step outside itself" to grasp the features of the object of knowledge. The rationalists' logical concepts and the empiricists' sense perceptions, in this context, are simply two major candidates for the task of identifying the source of cognitive representations of the object.

Kant, on the other hand, sees himself as resolving the conflict between rationalists and empiricists by introducing the object-constituting activity of the subject. In the Critique of Pure Reason, Kant argues that knowledge is rooted in the combined workings of the faculties of "receptivity" and "spontaneity." The former is the source of the raw materials of knowledge and the latter synthesizes them into the "formal structures" or objects of the experienced world.

> It is we therefore who carry the phenomena which we call nature, order and regularity, nay, we should never find them in nature, if we ourselves, or the nature of our mind, had not originally put them there. (Kant, 1966, p. A 123-127)

The important point for our purposes is that Kant (and later Piaget) effectively located knowledge in a new level of constructed reality -- the phenomenal realm. Roughly speaking, in that realm the possibility of knowledge is a function of the necessary *interaction* between subject and object (see Fabricius, 1983; Suchting, 1986).

The constructivist paradigm owes much to Kant's reaction to the epistemological turn Descartes initiated. In particular, constructivism shifts the ground of epistemology from the discovery of that which our experience is about to "the environment as we perceive it is our invention" (von Foerster, 1984, p. 42).

> The shift from knowledge-as-representation to knowledge-as-construction of reality implies an epistemological shift from the search after the "right method" for telling the true (really real) from the false (the apparently real) in what we observe, to the analysis of the operations allowing the observer to distinguish what he or she distinguishes and therefore to organize his or her own experience. (Chiari & Nuzzo, 1988, p. 92)

Ernst von Glasersfeld characterizes this paradigm as "radical" in that it avoids talk of a world existing independently of the observer -- a "world-in-itself." In what is clearly a pragmatic turn, radical constructivists maintain that knowledge, as a reflection or iconic representation of an observer-independent reality, must be replaced by knowledge as that which is in some sense "viable" in relation to the experiential world of the knowing subject (von Glasersfeld, 1984). Similarly, Maturana and Varela (1988) define cognition as "effective action" that permits a living being to continue its existence in a particular environment as it "brings forth its world."

The implication is that constructivists are giving up what philosophers have come to refer to as the "rational bridgehead" -- that allegedly unique common core of beliefs shared by all cultures. This, in turn, threatens to place us in a situation where our "justifications will begin to run in a circle and assume what they were meant to justify" (Barnes & Bloor, 1986, p. 27). However, the switch to knowledge-as-construction (or more generally, the pragmatic perspective on knowledge) offers a solution to questions surrounding self-reference and theory justification -- questions that are prone to circularity in the empiricist and rationalist traditions. The constructivist solution is to accept self-reference and other aspects of circularity as necessary features of a pragmatic theory of knowledge. As Heinz von Foerster writes:

> ...the observer, the observed phenomenon *and* the process of observation itself form a totality, which can be decomposed into its elements only on pain of absurd reifications... . (Quoted in Segal, 1986 p. xvi)

Maturana and Varela (1988) describe the unavoidable element of circularity in adopting a starting point for theory as a consequence of the "unbroken coincidence of our being, doing, and our knowing" (p. 24). Although the processes involved in our activities *constitute* our knowledge, we have no choice but to call on these very processes in order to talk about how we know. This is the case because what we do is inseparable from our experience of the world. As a result, constructivists, qua interactivists, are reluctant to engage in speculations about preconstituted subjects and objects as they may be independent of the active knowing process. In short, the observer has no operational basis upon which to make any claims about "things" (objects or relations) as if they existed independently of what the observer does.

However, in order to avoid the charge of solipsism or idealism, constructivists must address questions about the ultimate medium (for example, the source of Kuhn's (1970) paradigm-independent "stimuli") that serves as the setting for constructive activity. Here is Maturana's (1988) response:

> ...although it is an epistemological necessity to expect such a substratum, we constitutively cannot assert its existence through distinguishing it as a composite entity and thereby characterize it in terms of components and relations between components. In order to do so, we would have to describe it, that is, we would have to bring it forth in language and give it form in the domain of recursive consensual coordinations of actions in which we exist as human beings. However, to do so would be tantamount to characterizing the substratum in terms of entities (things, properties) that arise through languaging, and which, as consensual distinctions of consensual coordinations of actions, are constitutively not the substratum. (pp. 45-46)

"Languaging" something beyond the entities and properties of our experiential world (the "substratum," for example) is not possible. Although our epistemology "forces" us to talk about the substratum (that which "grounds" distinguishability) we must concomitantly hold that no "thing" exists in the substratum:

> All that we can say ontologically about the substratum that we need for epistemological reasons, is that it permits what it permits, and that it permits all the operational coherences that we bring forth in the happening of living as we exist in language. (Maturana, 1988, p. 47)

The potential explanatory power of a constructivist theory of knowledge derives from its analysis of the necessary connections between our forms of praxis and our philosophical ideas (see Johnson, 1989). That is, in their rejection of the traditional "problem of knowledge" and the associated difficulties of relating preconstituted "subjects" and "objects," constructivists have taken up the much more promising task of relating what we do to what we know. Indeed, in our efforts to understand the world there is always a close relationship between our purposes for studying the world and the very process of hypothesis formation (see Kitching, 1988).

Psychological Constructivism

For the most part, the constructivism advocated in this volume is more psychological than sociological in flavor. The reason for this cannot be attributed to the lack of sociological accounts that address the constructivist premises outlined above. Berger and Luckmann's (1967) *The Social Construction of Reality* could serve as a manifesto for these constructivists. Furthermore, many read Kuhn (1970) as both offering a constructivist account of the scientific enterprise and describing scientific ideas as existing and changing in social communities. Although these chapters give little attention to the social aspects of the objectification of mathematics, the accounts offered here are compatible with sociological interpretations in that they articulate types of cognitive processes that would be needed to ultimately explain sociological influences on knowledge construction. As Giere (1988) has suggested, a psychological account of knowledge construction should "provide a framework for understanding the role those [social] factors do play" (p. 16). Indeed Kuhn uses the notion of the gestalt switch to understand how individual scientists, driven by conflicts that arise from anomalies within a paradigm, suddenly come to view the world in a new way. Similarly, Berger and Luckmann describe "typificatory schemes" (for example, "man," "yuppie," "environmentalist") that pattern responses in face-to-face encounters. These typifications, however, can be modified when they become problematic during an interaction.

Schema Theory

The primary feature of cognition that constructivists invoke is its structure. They are particularly concerned with the role structured knowledge plays in determining even basic mental activities such as perception. We will refer collectively to these cognitive structures as schemata (the plural of schema). Other terms that fit under this generic descriptor include "frames," "scripts," "prototypes," and "plans" (see Nisbett & Ross, 1985). Because constructivists employ the notion of a schema in a variety of ways, we first provide a general description. We follow with a discussion of Piaget's more restrictive construct of the "scheme."

In this volume, Dubinsky (Chap. 9) offers a definition that captures the general notion of a schema: "a more or less coherent collection of cognitive objects and internal mental processes for manipulating these objects...[that are applied so as] to organize, or make sense of, a perceived problem situation." Neisser's (1976) definition adds to the organizing function of the schema the capacity to modify itself in response to perceptual information:

> [A schema is] that portion of the entire perceptual cycle which is internal to the perceiver, modifiable by experience, and somehow specific to what is being perceived. The schema accepts information as it becomes available at sensory surfaces and is changed by that information; it directs movements and exploratory activities that make more information available, by which it is further modified. (p. 54)

Alba and Hasher (1983) have challenged the usefulness of the schema as a theoretical construct partly on the basis of the ambiguity that is evident in these definitions. The difficulty of providing an adequate definition is perhaps one reason that schemata are typically introduced through the use of examples. These often consist of sentences or paragraphs that are difficult to comprehend, presumably because a suitable schema is not activated. Here is an example adapted from Bransford and McCarrell (1974):

After the seam split, the notes were sour.

Even though each phrase is reasonably intelligible, it is difficult to construct a coherent, overall interpretation of the sentence. The first phrase evokes ideas of clothing, or perhaps containers; the second conjures up music, maybe written sentiments. But it is difficult to locate in one's prior knowledge the theme that specifies the nature of the implied relationship between the two phrases. However, once the reader is given the word "bagpipe," the relation becomes clear. With this cue, a number of inferences are quickly made, inferences that are necessary for comprehension. One infers that the sour notes were *caused* by the split seam, on the one hand, and the continued effort to play the instrument on the other. One would expect that the restoration of the original tonal quality of the bagpipes awaits the repair of the seam. The original sentence contains none of this information. Rather it might be said to reside in the schema and is incorporated into the interpretation of the sentence.

A modern distinction borrowed from computing that relates to the function of a schema is that between "bottom-up" versus "top-down" processing. In the former, one begins with raw data and processes them to obtain higher-level meanings. In the latter, cognition is guided by higher-level, abstract information which selects and interprets lower-level data. In the top-down approach we form hypotheses or expectations of what we are experiencing before we have the data to "justify" such a hypothesis. Once a hypothesis is formulated, however, it searches for confirmatory data, choosing some data rather than others and making inferences in the absence of data. Nor is it accurate to say, as just implied, that the top-down process is initiated by the data. We are never without expectations that select a subset of information contained in the perceptual field. It is for these reasons that data are "theory-laden."

To summarize, schemata are the cognitive entities according to which constructivists' claim our knowledge of the world is organized. The idea that students must "construct their own understandings," follows from the fact that people are not bottom-up processors of information. What must not be overlooked, however, is that these constructed understandings are constrained in various ways and are not always successful in producing coherent interpretations of the world. In Popper's (1982) words, the external environment is "kickable" and is also capable of "kicking back." We discuss this feature in more detail below.

Piaget

The authors of these chapters have been especially influenced by Piaget's constructivist psychology. Among the reasons for this are that a) Piaget found a naturalistic metaphor for cognition in biology and, accordingly, his theory specifies the origins and evolution of higher cognitive functions; b) his theory offers an account of the development of mathematical thinking as arising out of the objectification of mental operations; and c) his method of actively probing children as they reason about and act on some aspect of their environment provides a welcome alternative to the Fisherian experimental techniques which have been paradigmatic in American psychology for most of this century.

The notion of "scheme" is central to Piaget's theory. A prototypical example of a scheme is the sucking behavior of an infant. At birth an infant will automatically suck objects that are placed into its mouth. Soon the infant learns to turn its head and open its mouth in response to objects that come into contact with the area near its mouth. The scheme in this example refers to whatever is in the newborn's cognitive system that organizes all of the actions required in order to permit the infant to turn toward and incorporate an object into its mouth for the purpose of sucking.

This example is particularly appropriate because it points to the biological metaphor which inspired Piaget's theory of mental activity and development. In the same way that living organisms convert objects in their environment into nutrients via the processes of digestion and absorption, the child's sucking scheme attempts to incorporate all kinds of objects into it. Thus a child is disposed to suck on virtually anything that it can get into its mouth. This process whereby a scheme is activated by, and directs actions upon, a particular object is therefore referred to by Piaget using the corresponding biological term "assimilation." Features of an object that are not relevant to the action triggered by the scheme are not "taken in" just as components of food consumed by an organism that it is not equipped to absorb are passed through the digestive tract and expelled. To an infant, a pen is something to suck, look at, or reach for. Features such as its use for making marks are not "processed" by the infant in the sense that there is no mental scheme that these features activate.

Unsuccessful Action

To this point, we have been describing a schema (using again the generic term) as a cognitive structure whose function is to organize or assimilate aspects of the environment. If an individual's current set of schemata were always successful in organizing its perceptions to bring about a desired state, there would be no further cognitive development. Development, in the constructivist view, is driven not by successful assimilation, but by the failure of assimilation to achieve the anticipated result.

Constructivists, adopting Piaget's terminology, refer to this capability of a schema to change in response to failure as "accommodation." Failures "disequilibrate" the cognitive system in much the same way that anomalies, in Kuhn's account, eventually throw a scientific community into a period of crisis. The cognitive system makes whatever changes are required to restore itself to a state of equilibrium, just as various forces ultimately lead to the adoption of a new "paradigm," restoring the scientific community to a state of "normal science."

When constructivists do invoke social constructs, it is often in reference to factors that propel the cognitive system into a state of disequilibrium. In particular, an exchange between two people is referred to in these chapters variously as a "didactical contract" (Confrey, chap. 8) or as involving a "negotiation" between participants that requires each to alter his or her perceptions in order to understand the other (Lewin, chap. 10; von Glasersfeld, chap. 4). The constructivist classroom is characterized not only by problems and teacher probes designed to facilitate cognitive change in the target domain, but by a style of conversation among students in which it is acceptable and desirable to question one another's beliefs, including those of the teacher. Constructivists generally agree that if this type of instruction is to take hold, common beliefs held by both teachers and students about their respective roles in the learning process will ultimately have to change. Paul Cobb has been especially outspoken about the need to bring about changes in the social climate of the classroom. This is not a straightforward task, however, but one that requires students and teachers "renegotiating the social norms that regulate classroom life" (Cobb, Wood, & Yackel, 1991).

As just noted, constructivists in general consider Kuhn's explanation of the succession of theories in the physical sciences to be compatible with the development of thought in individuals. However, a careful reading of these chapters would show them to be closer to Popper than to Kuhn on the issue of progress in development. Most of the chapters describe the development of mathematical thinking as the creation of progressively abstract constructions that do not so much replace as incorporate or coordinate prior conceptions. The result is a hierarchy of schemata, a notion of individual development that seems inconsistent with Kuhn's belief that successive theories in a scientific field are incommensurable. Popper's influence is clear in Bickhard's (chap. 5) characterization of the development of rationality as lying "not in accumulating positive content, nor in moving toward positive content, but instead in moving *away* from *error*" (Bickhard, chap. 5, p 73). Of course, these constructivists would disagree with Popper to the extent that he views falsifiability as a rational foundation for certainty rather than a method for achieving effective action. It might well be the constructivists' focus on pragmatism that allows them to steer a course between Kuhn and Popper, believing in the possibility that organisms and even societies may become increasingly more capable in their actions without, by implication, moving any closer to "absolute truth."

Foundations Revisited

We began by discussing the traditional "problem of knowledge" in the context of a radical subject-object dichotomy. We then suggested that Kant effectively completes the "epistemological turn" with his emphasis on the object-constituting activity of the subject. Kuhn's "post-empiricist" theory of science represents further development along these lines in offering a radical alternative to the mainstream "rational reconstruction" of the proper criteria and standards of science. Kuhn's criticisms strike another blow to the attempt to ground knowledge in a foundation that would permit scientific theories (or knowledge in general) to be "justified" according to universally agreed-upon criteria:

> [There is] no neutral algorithm for theory choice, no systematic decision procedure which, properly applied, must lead each individual in the group to the same decision. (Kuhn, 1970)

The power of placing the constructivist response to the "problem of knowledge" in the larger context of the grand Cartesian "either/or" is apparent when we consider one of the main reasons why Kuhn (and others) meet with such great opposition. Kuhn's critics argue as if we must choose between the alternatives of objectivism or some form of radical subjectivism. On the post-Cartesian view, however, the anarchistic element in Feyerabend's (1975) Against Method, the relativism of Kuhn, the condemnation of "instant rationality" in Lakatos (1970), as well as the radical constructivism of Maturana and von Glasersfeld all can be seen as instances where a theorist is not taking sides with the Cartesian either/or. They maintain, rather, that the rejection of foundationalism in all its forms, far from signalling the irrationality of science or "Reason," merely focuses our attention on the evolving and socially situated character of rationality and "truth." This reading of these authors' works does justice to their stated, pragmatic goals, while keeping them free of the more traditional problems associated with attempts to give priority to one or the other pole of the Cartesian duality.

Overview of Chapters

These chapters reflect the range of programs that comprise the constructivist movement in mathematics education. These can be broadly characterized as questions involving a) foundational issues in constructivist epistemology; b) the nature of knowledge representation and mechanisms of representation change; c) characterizations of mathematical knowledge at various levels of expertise; and d) the development of teaching practices and research methodologies that are consistent with the constructivist paradigm. Although most of the chapters deal with aspects of several of these questions, we have attempted to place each chapter under the most suitable heading.

Foundational Issues

Both Steffe (chap. 3) and Bickhard (chap. 2) address a logical challenge to empiricism which turns out to be a challenge as well to constructivism. If the environment cannot impress itself directly into the mind, then how does learning occur? If one begins with the view that learning is a form of induction that involves abstracting patterns from observations, then those patterns could not be perceived unless they were already in the mind. Thus, nothing new could be learned through induction. Such arguments seem to support an extreme form of innatism.

In analyzing Fodor's innatist response to this problem, Bickhard (chap. 2) uncovers what he suggests is a false premise in Fodor's argument, one that nevertheless is shared by most cognitive theorists. This is the notion that knowledge, qua representations, involves the pairing of some mental "element" with an existing element that already has an established meaning. On the basis of this view of representations, new meanings can never arise. Steffe (chap. 3) approaches the same problem through a case study which he offers as an example of a type of learning that is neither innate nor based on inductive inferences. He then proceeds to distinguish "engendering" and "metamorphic" accommodations in terms of which he reformulates a type of induction (abduction) that he presents as a solution to the "learning paradox."

General Mechanisms of Constructivist Learning

Reflective abstraction. Piaget's construct of reflective abstraction has played an important role in constructivist treatments of the development of mathematical thought. It has served as a means of describing the process of constructing increasingly abstract concepts from purely mental objects. Most of the chapters make reference to this process. von Glasersfeld (chap. 4) clarifies Piaget's distinctions among three types of reflective abstraction and focuses especially on one variety that requires conscious awareness. von Glasersfeld (chap. 4) characterizes mathematical thinking as consisting primarily of abstractions derived from conscious thinking about the operations of thought. Furthermore, the capability of symbols to "point to" such abstractions permits operations to be "re-presented" in the mind in summary or incomplete form rather than having to be enacted in toto.

Knowledge hierarchies. In his chapter on rationality, Bickhard (chap. 5) claims that reflective thought is a product of the hierarchical development of knowledge. In developing this idea, he shows principles of rationality (including the idea of necessity) to be emergent features of cognitive activity that seeks to minimize error. In their chapter, Kieren and Pirie (chap. 6) explore the power of viewing mathematical thought as recursive. In their examples, mathematical operations at one level in a hierarchy serve as inputs for a higher level, yielding a self-referencing set of schemata that can be quickly and flexibly traversed.

Repeated experience. Cooper's (chap. 7) data suggest that repeated experience within a specified domain can lead to knowledge reorganization. This finding represents a challenge to constructivists who would not be able to account for learning in which there was no disequilibration. Cooper (chap. 7) is quick to dismiss the explanation, based on associationist models, that such experience results in the strengthening of stimulus-response bonds. In contrast, he likens the outcome of repeated experience to the construction of a richly, interconnected cognitive map that is built up from the experience of navigating city streets. Here he discusses the role of inconsistencies that can lead to cognitive re-organization. Indeed, constructing such a "map" may require a class of perturbations discussed in Steffe (chap. 3) which result from reorganizing only *part* of a well-defined structure.

Describing Mathematical Schemata

At the heart of the constructivist program are studies of how people think about various mathematical entities and processes. Confrey (chap. 8) demonstrates the effective use of interviews in exploring student conceptions in a particular domain. Her analysis of one student's protocols avoids any simple characterization of the student's view of exponents, but rather emphasizes the incompleteness, complexity, and tortuous evolution of his thinking. This student's struggle with creating a consistent and meaningful structure for exponents should serve as a reminder that the efficiently operating hierarchies described above result from the resolution of innumerable contradictions.

Educational Applications

Curriculum design. Traditionally, educators design curriculum on the basis of a task analysis in which a particular domain is broken down by "content experts" into its rudimentary components. These components are then arranged in a "logical" order in the curriculum according to their relative complexity and interdependency. If these experts have earned their pay, students proceeding attentively through the curriculum will not encounter novel tasks for which they lack the requisite skills.

The haste with which constructivists could articulate reasons for rejecting this approach would stand in contrast to the cautiously worded principles or heuristics of the alternative they might offer. It is therefore to his credit that Dubinsky (chap. 9) attempts to identify and teach general constructive activities that are necessary for understanding complex mathematical and logical entities such as functions, proof by induction, and predicate calculus. His instruction is based on "genetic decompositions" which function as task analyses but arise out of a dialectic between his own understanding of the target domain and his evolving conceptualization of his students' understandings.

Student motivation. Motivating students in traditional classrooms often means getting them to listen. Teachers traditionally achieve this with skilled performances or, more typically, with the threat of a quiz. The requirement that students construct meanings for themselves makes the constructivist teacher's job of motivating students both more critical and complex. In his chapter, Lewin (chap. 10) describes a teaching approach in which students become actively engaged in interpreting the classics. Texts generate both questions that matter and problems that require dialogue. They become "vehicles" through which their authors converse with (rather than simply talk to) their reader. Lewin (chap. 10) invites us to consider the body of mathematics as an artifact, the examination of which may provide insights into how we and others think.

Constructive computing. The computer has something to offer every teacher, regardless of his or her philosophy of education. For example, the computer has been used as a high-tech version of Skinner's "teaching machine," providing immediate feedback and "branching" students to different parts of the program based on individual performance. Hatfield (chap. 11) describes what a constructivist educator might see when looking at a computer: students inventing their own algorithms for solving problems and designing mathematical simulations of physical systems. He pays particular attention to those activities that allow students to invest themselves such that they might be said to have had a "mathematical experience."

2 The Import of Fodor's Anti-Constructivist Argument

Mark H. Bickhard

Fodor argues that the construction of genuinely novel concepts is impossible and, therefore, that all basic concepts available to human beings are already present as an innate endowment (1975, 1981). This radical innatism - along with related conclusions such as an innate modularity of available representations and a corresponding innate limitation in the potential knowledge that human beings might be capable of (1983) - has been seen by many as a reductio ad absurdum of Fodor's position, and his arguments have consequently been dismissed. I will argue that Fodor's arguments deserve much more careful attention than that: in particular, his arguments *are* a reductio of one of his essential presuppositions, but it happens to be a presupposition that he shares with virtually all of psychology and philosophy. Fodor's conclusions, then, are reductios of the major portion of contemporary studies of cognition and epistemology (Campbell and Bickhard, 1987). Furthermore, even when the critical presupposition is isolated, it is difficult to construct a genuine alternative. Most attempts at correcting any part of the logical difficulties involved have inadvertently presupposed the pernicious premise elsewhere in the system (Bickhard, 1980a, 1982, 1987).

Fodor's Argument

Fodor's arguments are usually stated in terms of theories of learning or concept learning and yield the conclusion that there are in fact no such theories: what claim to be theories of learning are really theories of belief fixation (Fodor, 1975, 1981; Piattelli-Palmarini, 1980). Fodor's basic point is that all such theories require that concepts or hypotheses be already present before learning can begin; 'learning' is then limited to establishing - or failing to establish - belief in a preexisting concept or hypothesis. The new concept, the new representation, must be already constructed before learning begins. (This is just Popper's argument against the logical possibility of passive induction [Popper, 1959, 1965, 1972, 1985], but Fodor seems unaware of that, or unwilling to acknowledge it. He even persists in calling such hypothesis testing "induction.")

According to this argument, then, learning cannot create new representations. The representations that learning is supposedly concerned with arise not from learning itself, but from constructions of structures of prior representations. This is a combinatoric type of model in which new combinations of already present representations may be assembled and then tested against experience via "induction." Such a view, however, requires that there be other representations already present that can participate in the combinations, and although some of those may themselves be products of still earlier combinatory constructions, there must be some basic ground of representations that suffices for all possible such combinations. The alternative is an infinite regress of combinations of ever more primitive representations. Because this would be a regress of actualities - of actual representations, supposedly in the mind - it is unacceptable.

By assumption, the basic representations cannot themselves be decomposed into combinations of lower order representations, but no models of learning or development provide any way for such fundamental representations to be created. Therefore, Fodor reasons, these basic representations must be innate, i.e., provided in the genome. The conclusion is that some basic set of representations, combinatorically adequate to all possible human cognitions, must be innately present.

Fodor presents a number of related arguments that I will not pursue directly, but that depend in part on the innatist conclusion of the above argument, so I would like to indicate the nature of that dependence. First, because there is little evidence for any combinatorial complexity in studies of lexical comprehension, Fodor concludes that any combinations from innate primitives that do exist are relatively shallow. The result is that the resources of the innate base must be very strong indeed, because even the power of unrestricted combinatorics seems to be unavailable to the constructive processes founded upon it (Fodor, 1981). Second, along with other arguments, the assumption of an innate base for all possible human representation yields as a plausible conclusion that human representational (and inferential) capacities are clustered in innate modules (Fodor, 1983). Third, the representational power offered by such an innate base of representational atoms is at the same time a restriction of power: anything that would require a representation that can*not*, in principle, be combinatorically constructed from those atoms is intrinsically beyond human capacities to represent or understand (Fodor, 1981, 1983). I mention these subsidiary arguments not in order to address them in detail, but rather to indicate their relationship to the basic innatist argument and, therefore, to indicate the sense in which they, too, are undermined should Fodor's basic argument prove to be flawed.

Reconstructing the Argument

Stating the argument in terms of learning or development is, I suggest, somewhat misleading. It is misleading because it directs attention to theories of learning or development as being the source of problems (should the argument be taken as revealing a problem, rather than just as explicating the way of the world). The more fundamental source of the impasse which Fodor's argument reaches, however, is not located in theories of learning or development per se, but instead in the theories of representation that are presupposed by those approaches to learning and development - and by virtually all the rest of cognitive and epistemological studies as well.

The core of the difficulty is that the only way that these notions of representation allow for new representations to be constructed is as combinations of elements that are themselves *already representations*. The rules of combination may vary enormously in differing models, as may the presumed base of representational atoms (should the model be so explicit as to acknowledge and address the necessity for such a representational periodic table), but the presupposition of "representations only from representations" holds universally. After all, "out of nothing nothing comes" (Fodor, 1975, p. 59).

But this implies that one must already have representations in order to construct representations, and that implies, in turn, that it is impossible for representations to be constructed, to come into being, out of foundations that are not themselves representational. It follows, then, that it is impossible for representations to emerge out of non-representational phenomena or processes. It is this impossibility that is at the heart of Fodor's anticonstructivism: learning and development are incapable of creating *emergent* representations; they can only create new combinations of old representations. The origin of this seeming impossibility, however, is in the notions of representation involved, not directly in the models of learning or development.

In effect, then, Fodor's arguments have turned on the inability of contemporary studies of cognition and epistemology to account for the *emergent nature* of representation. Such studies account at best for the combinatoric nature of *non*-emergent representation, but they are thereby vulnerable to Fodor's challenge.

If representations cannot emerge, however, then they cannot come into being at all. A narrow focus on this point yields Fodor's innatism: neither learning nor development, as currently understood, can construct emergent representation; therefore the basic representational atoms must be already present genetically. Unfortunately, this conclusion does not follow. If representation cannot emerge, then it cannot emerge in evolution any more than it can in development. The problem is logical in nature and is not specific to the individual. Conversely, if some way were posited in which evolution *could* yield emergent representation, then there is no a priori reason why that emergence would not be just as available in the development of the individual. Fodor's innatism, then, simply misses the basic issue. If representation cannot emerge, then it is impossible for it to exist, and evolution is in no better position in this respect than is individual development; on the other hand, if representation *can* emerge, then there is something wrong with the models of learning and development that cannot account for that emergence. When those models are corrected, that emergence should be as available to the individual as to evolution. In either case, Fodor's strong innatism does not follow. (In this final innatist conclusion, then, Fodor's reasoning is *not* valid.)

Clearly, representations do exist, and, therefore, representation *can* emerge from non-representational phenomena. Therefore, equally clearly, there *is* something wrong with contemporary models of representation that make that emergence logically impossible. Fodor's anticonstructivism is a reductio of models of representation that make the emergence of representation logically impossible, but, ipso facto, Fodor's innatism is, therefore, not a viable solution to the problem.

What's Wrong with Contemporary Models of Representation?

Fodor's anticonstructivism turns on a basic flaw in contemporary models of representation: the inability to account for the emergence of representation. This is an abstract indictment, however, which still leaves open the question of *why* contemporary models are subject to this charge. In this section, I will argue that the root problem is in the assumption or presupposition (it's not always explicit) that representations are constituted as some sort of encodings (Campbell and Bickhard, 1987).

Paradigmatic encodings are such schemes as Morse Code, in which, for example, "..." stands for "S." Morse code and other similar codes, such as computer codes, security ciphers and codes, and so on, are unproblematic with respect to the issues at hand, but the very nature of their "unproblematicness" reveals some of the deep sources of difficulties in more epistemologically ambitious encodingisms. The encoding, "..." say, in such a scheme is explicitly paired with another representation that it is to be taken as standing in for-"S" in this case. The encoding correspondence is explicitly defined, and that which the correspondence is with-"S"-is itself an unproblematic representation of the sound /s/. All actual encodings are constituted by explicit known correspondences with known defining representations. It is only when it is assumed or presupposed that such a model could hold for epistemology in general that fatal difficulties emerge.

Encodings, fundamentally, are stand-in representations. In Morse code, "..." stands in for "S" and picks up the same representational content that "S" carries, namely /s/. "S" provides that content by already carrying it prior to the definition of Morse code, and "..." is established within Morse code precisely by being paired with "S" as a stand-in. Note that "..." obtains its representational content from "S," and "S" is presumed to *already* carry *its* representational content. Nowhere in the nature of an encoding is there any possibility of a *new* representational content being established, except in the purely combinatoric sense of an encoding element being taken to stand in for some combination of previously established elements. Encodingisms, then, involve precisely the inability to address the emergence of representational content upon which Fodor's arguments turn.

For another perspective on this issue, consider the definition of an encoding as being an element with a known representational content. This definition is, in fact, equivalent to the 'stand-in' definition just mentioned. In one direction of implication, a stand-in definition *provides* a known representational content to the defined element, thereby *making* it an element with a known representational content. In the other direction, any specification of representational content for the definition of an encoding element requires some other representation(s) which can provide or specify that content, and the new element then becomes a stand-in for whatever representation(s) provided that content. This definition, however, makes it clear that any representation that is constituted as a representation by virtue of the representational content which it carries is intrinsically an encoding. Any representation defined as a representation in terms of what it represents is an encoding.

But a representation defined as such in terms of what it represents *cannot* address the issue of representational emergence; in such a scheme, representational content must already be available before any encoding representation can be defined. There is no starting point, no origin, for the definitional process. If it is presupposed that encodings *can* somehow create new representational content - can provide the foundations for their own definitions - a logical incoherence results. In particular, any finite number of stand-in definitions is unproblematic: they are "simply" passing on already available representational contents. But when we address the foundational representations upon which others are to be defined, impossibilities emerge. If such foundational representations are presumed to be encodings and if we ask how those foundational encodings are provided with representational content, no answer is possible. Put another way, if we ask how an epistemic agent could possibly know what a foundational encoding is *supposed* to represent, no answer is possible. The point is made simply by noting that, on one hand, if the foundational encoding is provided with content in terms of any other representations, then it is not foundational, contrary to hypothesis.

On the other hand, if such a foundational element, say "X," is presumed to provide its *own* content, we have ' "X" represents whatever "X" represents' or '"X" stands in for "X." ' In neither case is any representational content established, and, therefore, in neither case is "X" established as an encoding at all. Consequently, the notion of a foundational encoding is logically incoherent. Presupposing such a foundation leads to the incoherent conclusion of an encoding element with no representational content.

This incoherence is, in fact, just the *logical* aspect of the inability of encoding models to address the emergence of representational content. Such emergence, if it is going to occur anywhere, will have to occur at the foundations of a strict encoding scheme, and, in fact, it *does* have to occur *somewhere* if any version of encodingism is to "get off the ground." But any actual encoding scheme is only capable of providing and carrying representational content, not creating it. Consequently, the presuppositions of an encodingism necessarily include the incoherent presupposition that foundational encoding elements are possible, that encodings can, by themselves, do something that they cannot do: create new representational content.

Why Should We Care?

One deeply important consequence of this incoherence of encodingisms, of this inability of encodings to account for new representational content, is that encodings *cannot* be the answer in any case where *new* representational content is required. Simply stated, encodings cannot cross epistemic boundaries. Encodings cannot provide basic representations for basic domains of knowledge, they can only change the form of representations in or across *already represented* domains of knowledge. In particular, encodings cannot provide the basic representations for the mind about its environment - whether perceptual or otherwise (Bickhard and Richie, 1983). Similarly, they cannot provide a basic representational access from the world - language, say - *into* a mind (Bickhard, 1980a, 1987). In neither case can encodings create the foundational representational contents required to start the process of representing. *Encodingisms, in other words, are logically incapable of ANY of the basic epistemological tasks for which they are standardly proposed.*

That fundamental inadequacy is the logical fulcrum for Fodor's arguments, and, consequently, we ignore those arguments at our peril. They point to a deep flaw in contemporary epistemology that is not at all easy to avoid - virtually *all* available models of representation are encodingisms (Bickhard, 1980a; Bickhard, 1982; Bickhard and Richie, 1983; Campbell and Bickhard, 1986; Campbell and Bickhard, 1987; Bickhard, 1988; Bickhard and Campbell, 1989; Bickhard and Terveen, 1990). The proper conclusion from Fodor's arguments, however, is not one of a radical nativism. Nativism and anticonstructivism, in fact, manifest only partial and distorted insights into the fundamental problem. It is, rather, "that the argument has to be wrong, that *a nativism pushed to that point becomes unsupportable, that something important must have been left aside*. What I think it shows is really not so much an a priori argument for nativism as that *there must be some notion of learning that is so incredibly different from the one we have imagined* that we don't even know what it would be like as things now stand" (Fodor in Piattelli-Palmarini, 1980, p. 269). Actually, as mentioned, not quite, the fundamental problem lies in our notions of representation, not in our notions of learning.

In-principle arguments concerning the inadequacy of a modeling approach, such as the argument concerning encodingism, can at best be disquieting unless there is a plausible alternative. Without an alternative, it is difficult to know what to try to correct and what implications such a correction might have. Even at the level of disquiet, however, I wish to emphasize that Fodor's arguments, along with, most importantly, the presuppositions upon which they turn, should be cause for great unease. The basic encodingist assumption of *all* contemporary approaches to cognition is fundamentally incapable of solving *any* of the basic epistemological problems with which we are concerned. The resulting anticonstructivism, nativism, intrinsic limitations to knowledge, and so on, are all very much secondary to that basic difficulty. Furthermore, with so foundational a flaw in our most central presuppositions, it is not at all clear what sorts of guidance from the resultant models we can trust in perception, cognition, language, learning, development, education, instruction, or any other domain involving representation. I do not wish to leave this discussion only at the level of disquiet, however. There is an alternative approach to the nature of representation, called interactivism, and I will give a brief and limited introduction to it here (Bickhard, 1980a; Bickhard and Richie, 1983; Campbell and Bickhard, 1986).

Interactivism: Outline of a Solution

The incoherence of encodingism originates in a basic circularity in the notion that an element can be an encoding only if it carries representational content, yet encodings are taken to be the only source of representational content. When pressed at the level of foundations, then, encodings must provide their own contents in order to be encodings at all, but in order to provide such contents, they would already have to be encodings. There is no way out of the circle and, thus, no way to avoid the incoherence.

Interactivism avoids this circularity and resulting incoherence by separating the condition of being a representation from the condition of carrying representational content. It becomes possible for representational elements to come into being *without carrying any representational content* and only subsequently for representational content to be given to them. This breaks the circularity and thus avoids the incoherence. This abolishes the requirement that we must already have representations in order to get representations.

It does so, however, only by providing a definition of representation that is more primitive than that of "carrying representational content," but that can, once extant, address the problem of representational content. The problem for interactivism as an alternative, then, is twofold: first, provide a notion of representation that does not require the carrying of representational content; and second, provide a notion of representational content that can be constructed subsequent to the construction of a basic representation and that can be carried by it.

The foundational conception here is that of interactive differentiation. First, an *epistemic* system is argued to intrinsically be an *interactive* system (Bickhard, 1980a, 1988; Bickhard and Richie, 1983; Bickhard and Campbell, 1989). Knowing is successful interacting, and knowledge is the capability for knowing. System processes of knowing, of successful interactions, provide the non-representational foundations for explicating representation. The internal course of process in an epistemic system will depend jointly on the organization of the system and on the environment engaged in the interaction. Consequently, the final state in which that system (or subsystem) is left when the interaction is completed will also depend jointly on the system and on the environment. Each possible final state of such an interactive system or subsystem will correspond to the class of possible environments defined by the condition that each environment in the class would yield that final state in an interaction. Each final state, then, *differentiates* its class of environments from those environments that would yield some different final state. This notion of interactive differentiation is more primitive than the notion of carrying representational content. A final state differentiates its class of possible environments from all others, but *it doesn't represent anything at all about those environments except (implicitly) that they yield this final state rather than some alternative final state*. There is no representational content here, only a functional and implicit differentiation.

Interactive differentiation, then, does provide a notion of representation that is sufficiently primitive as to not require representational content. The next step is to model how representational content could be provided for those environments that are "contentlessly" differentiated. The core intuition for this step is to realize that environmental differentiations may be useful for the system in further differentiating its own internal activities. Environments of the sort that yield final state "A" may be amenable to strategy "S22," while environments of the sort that yield final state "B" may be more appropriate to strategy "S87." But such knowledge implicit in the functioning of the system *is* representational content. If the system can select the appropriate strategy given an environmental differentiation, that *constitutes* representation of a property or properties of that class of environments. It constitutes representation of the implicit predication that "final-state-A type environments" are also "appropriate-to-strategy-S22 type environments." In this way, a final state can come to acquire representational content, perhaps vast content in complex organizations and webs of such selectional properties. Further, once given representational content, an interactive representation can serve to provide representational content for the definition of a *derivative* encoding; encodings can be defined as stand-ins for interactive representations.

Interactivism's claim to be an alternative to encodingism and one that avoids encodingism's tangle of incoherences and absurd consequences is a programmatic claim. It is programmatic in the sense that interactivism is a general approach to epistemological phenomena whose potential can ultimately be demonstrated only by constructing useful particular models within that general interactive framework. There will be little programmatic elaboration here, so interactivism will be presented as a general programmatic approach based on in-principle arguments. More extensive elaborations of interactivism can be found in Bickhard (1980a), Bickhard (1980b), Bickhard and Richie (1983), Campbell and Bickhard (1986), and Bickhard (1987).

There are two consequences of interactivism, however, that I will briefly present because they are responses to prima facie serious challenges to interactivism's general programmatic claims. The first challenge is simply to turn the root of the encodingism incoherence problem back to interactivism: How can interactivism account for the emergence of novel representation, of representation constructed out of non-representational foundations? The general form of the response is that interactivism logically forces a *constructivism*, and a constructivism that does in fact provide a solution to the problem of representational emergence. The second challenge is to the general representational adequacy of interactivism. Interactive representation emerges as a functional aspect of goal-directed interactive systems. Even if the direct claims of interactivism are taken at face value, it might still seem that interactivism could at best account for representation of physical environments. Interactive representation requires interaction, and interaction requires something to interact with, and (so the challenge goes) it is not clear what the interactive environments could be for abstract knowledge such as that of mathematics. The general form of the response to this challenge is to show that interactivism yields a hierarchy of levels of potential knowledge in which each level is of increasing abstraction relative to the levels below.

I turn first to the interactive account of emergent representation. The temptation within encoding models is to try to account for the creation of representations via the passive creation of element-by-element and structural correspondences with things and patterns in the world. Encounters with the world will result in things, events, and patterns in the world impressing themselves into a receptive mind, thereby creating encoding correspondences, the presumed impressions, with those things, events, and patterns. The classic version of this is the tabula rasa, the blank waxed slate, but contemporary notions of transduction and induction are simply more sophisticated versions of the same idea (Bickhard and Richie, 1983). Unfortunately, even if the basic passivity of mind in these models is accepted and even if the passive creation of such correspondences is also accepted, an encoding is constituted as a *known* correspondence, known both in terms of the fact of such a correspondence and in terms of what the correspondence is with. The incoherence argument shows that the origin of that knowledge *about* the correspondences is precisely the fatal inadequacy of encodingism. Factual correspondences are easy to find, they occur in physics all the time; it is knowledge that is difficult to account for.

Interactivism, in contrast, does not offer the same temptation. Interactive representation is not a matter of correspondence relationships at all but, rather, an emergent of interactive functional relationships. There are *no* particular element-by-element or structural relationships between an interactive differentiating system and the environments which it serves to differentiate. There are only open, functional differentiating relationships between a system and its environment. The system differentiates, but it does not correspond.

The final states might be said to correspond to their differentiated classes of potential environments, but this can only be specified from the perspective of an external observer of the system who independently *already has* representations of *both* the system's final states *and* of the environments that are differentiated, and who can therefore *define* the correspondences between them. Such correspondences can be defined within the *observer's* perspective on the open system differentiations, but the *system itself* has no representations of such correspondences, nor even necessarily of their existence, and they can play no direct epistemic role for the system (cf. Maturana and Varela, 1980, 1987; see Bickhard and Terveen, 1990). The fact, however, that system differentiations can be described by an external observer in terms of correspondences is one more reason why encodingism has enjoyed such credence in spite of its logical difficulties.

Interactive representation is a matter of a system's functional relationship with its environment, not a matter of structural relationship. Any notion of the environment creating a structural relationship, therefore, is irrelevant. Furthermore, there is no possibility that the environment can passively impress or create a *system*. Interactive representation, then, can be created only by the *internal constructions* of the system itself. The system must try out new system organizations and test them against the environment, a variation and selection constructivism.

Still further, those internal system constructions are of system organization, *not of representations*. The construction of interactive representations is emergent in the construction of system organization and *does not occur* (except with respect to secondary encodings) in terms of combinations of already available representations. A solution to the problem of representational emergence, then, is intrinsic in the nature of interactive constructivism. System constructions construct new system organization, and representation emerges as a functional aspect of system functioning, so already existing representations are not required.

I turn now to the second challenge, concerning the interactive account of abstract knowledge. The basic intuition for the problem of abstract knowledge is fundamentally Piagetian: the properties and activities of an interactive system are more abstract than the environmental properties being represented by that system. We begin, then, with a first-level system that interacts with and represents its environment. This system, in turn, will have properties that could be represented by a second-level system interacting with the first level system. The first-level system would be the interactive environment for the second-level system. The second-level system, in turn, has still more abstract properties that could be represented from a third level, and so on indefinitely. Interactivism, in this manner, generates an unbounded hierarchy of potential levels of representation, of potential interactive knowing, of ever-increasing abstraction. Interactivism, then, does not face an impasse with respect to abstract knowledge, but instead has unboundedly rich resources for addressing abstractions.

Furthermore, these levels of potential knowledge are not merely an ad hoc solution to the problem of abstraction, but instead have many other properties that connect with other phenomena in psychology and epistemology. For example, constructivism forces the levels of knowing to be ascended one at a time, in sequence, from the first level upward. It is impossible for any system to exist, to be constructed, at a given level unless there is already something to be interacted with at the level below. The hierarchy of levels of potential knowing, then, forces an invariant sequence of stages of constructive development. These stages, in turn, account for the major qualitative shifts found in development (Campbell and Bickhard, 1986). The levels, in other words, do far more than just account for abstract knowledge.

For another example, I argue that the *constructive* relationship between one level and the next higher level is essentially that of Piaget's notion of reflective abstraction (Campbell and Bickhard, 1986), while the *epistemic* relationship between one level and the level below is that of reflective consciousness (Bickhard, 1980a, 1980b; Campbell and Bickhard, 1986). Interactivism, then, by forcing constructivism and levels of potential knowing, accounts in a natural way for the related phenomena of reflective abstraction and reflective consciousness. In a similar manner, many other phenomena of psychology and epistemology emerge via natural elaborations of the basic interactive approach.

Interactivism, then, does not fall prey to the incoherences of encodingism, but rather offers a rich alternative perspective on the phenomena that are traditionally conceded to encodingism. In the context of this discussion, the critical point is that interactivism does offer a way out of the impossibilities and absurdities presupposed by and revealed in Fodor's arguments.

Conclusions

Fodor's arguments appear to reach absurd and unacceptable conclusions. On that basis, it can seem all too easy to dismiss them. I have argued that Fodor's conclusions *are* absurd and unacceptable, but that his arguments are founded on a fundamental flaw in contemporary approaches to the nature of representation. If that is the case, then it is logically inconsistent, though not trivially so, to both reject Fodor's arguments and accept contemporary approaches to representation. The arguments follow directly from the encodingist approaches, and encodingism is ubiquitous in those contemporary approaches. Still further, encodingism is extremely difficult to abandon, both because there is no readily available alternative and because many positions have deep, non-obvious logical commitments to an encodingism. It is not easy to uncover and replace implicit encodingisms (Bickhard and Richie, 1983). I have briefly outlined an alternative to encodingism called interactivism. Interactivism is not susceptible to the impossibilities and absurdities of encodingism. The replacement of encodingism and its equivalences by interactive models will be a slow, programmatic process, but encodingism itself is epistemologically impotent and incoherent and, therefore, *must* be replaced.

3 The Learning Paradox: A Plausible Counterexample

Leslie P. Steffe

In the summer of 1985, Carl Bereiter published an article in the Review of Educational Research titled *Toward a Solution of the Learning Paradox*. Ever since, it has been my intention to provide a counterexample to the paradox. Fodor (1980b), who is credited by Bereiter as clearly stating the learning paradox, views learning as being necessarily *inductive*. "Let's assume, once again, that learning is a matter of inductive inference, that is, a process of hypothesis formation[1] and confirmation" (p. 148). Given his view of learning, Fodor states the learning paradox in the following way.

> There literally isn't such a thing as the notion of learning a conceptual system richer than the one that one already has; we simply have no idea of what it would be like to get from a conceptually impoverished to a conceptually richer system by anything like a process of learning. (p. 149)

Moore simply, "the paradox is that if one tries to account for learning by means of mental actions carried out by the learner, then it is necessary to attribute to the learner a prior cognitive structure that is as advanced or complex as the one to be acquired" (Bereiter, 1985, p. 202). In other words, if learning is viewed as inductive inference, mathematics cannot be learned but has to be conceived of as a mental entity that is programmed to unfold over time (Gardner, 1980).

A Guiding Analogy

Counterexamples have played an important role in the development of mathematics and it is possible they could play an analogous role in the development of a psychology of mathematics learning. For example, in the history of mathematics, no one succeeded in establishing that the parallel postulate was a consequence of the other postulates of synthetic Euclidean geometry (Moise, 1963, pp. 130-31). The attempts to establish independence of the postulates led to the development of non-Euclidean geometries in which all postulates other than the parallel postulate held along with a negation of the parallel postulate. In attempts to prove the parallel postulate from the other postulates of Euclidean geometry, the creative work stemmed from denying the postulate. The non-Euclidean geometries that were developed were themselves based on different models of lines and planes, the undefined terms of the postulates. Spherical geometry is based on eliminating lines through a given point that do not intersect a given line. The development of hyperbolic geometry, the other non-Euclidean geometry, had its impetus in the second possible denial of the parallel postulate, where at least one line passes through a given point that is parallel to a given line (Eves & Newsom, 1958, p. 69).

It is not possible to always use a model of learning belonging exclusively to one learning theory or to another like it is to use a model of lines and planes belonging to a particular geometry. This has been lucidly demonstrated in the debates between the associationists and the field theorists during the middle of this century (Henry, 1942) and tempers my expectation for a surge in the development of a psychology of mathematics learning. Nevertheless, there is a useful analogy between how learning is reformulated and the historical work in non-Euclidean geometry.

One interpretation of denying Fodor's definition of learning simply eliminates learning as inductive inference. This would be compatible with spherical geometry. The other interpretation, and the course I intend to follow, asserts that learning is at least inductive inference. But, the model of inductive inference that is formulated differs from the model as explained by Fodor.

An Example of the Learning Paradox

Before reformulating the meaning of learning, I turn to a discussion of an alleged example of the learning paradox explained by Bereiter because it is essential in understanding my counterexample. Children have been found to make a transition, without any instruction, from a crude addition algorithm to a more efficient one (Groen & Resnick, 1977). Using the crude algorithm, children find the sum of four and three by counting out four blocks, counting out three blocks, and then counting them all. Using the more advanced algorithm, children start counting from four and count on three more (Steffe, Hirstein, & Spikes, 1975, pp. 95 ff). Bereiter claims that we run right into the learning paradox when attempting to explain how children learn to count on.

> To attribute this step to some kind of insight--to "realizing" or "seeing" that counting is unnecessary because the resulting number is already given--is, of course, to tumble right into the learning paradox. For such an insight

> presupposes an understanding of the more sophisticated procedure in advance of discovering it. (p. 203)

Bereiter goes on to recount attempts by Klahr and Wallace (1976) to explain the phenomenon using information processing principles--inherent self-modifying properties of a cognitive system like consistency detection and redundancy elimination.

> According to the principles, consistency detection would lead to the knowledge that starting with n objects and counting them out always results in the number n, whereupon the counting operation would be eliminated as redundant. (Bereiter, 1985 p. 203)

These two principles appear to be less complex than the learning they are supposed to explain, but Bereiter believes that elaborate innate systems for re-presenting numerical information would have to be attributed to the child if the principles applied. This would lead to what Chomsky and Fodor insist is the necessary alternative to constructivism, an alternative that leads to the belief that number is innate (Chomsky, 1980b; Fodor, 1980b).

> If one rejects the hypothesis of innate mathematical knowledge, then one is faced with the learning paradox: How can the child's recognition of a redundancy in the addition procedure be explained without attributing to the child prior knowledge of cardinality? (Bereiter, p. 20)

I certainly agree with Bereiter[2] that if one rejects the hypothesis of innate mathematical knowledge and if one views learning only as inductive inference, it is impossible to avoid the learning paradox in this example.

A Weak Form of the Innatist Hypothesis

If, as Chomsky (1980b) claims, there is an intrinsic system that specifies the phonetic, syntactic, and semantic properties of an infinite class of potential sentences, then we should see a corresponding development of the verbal number sequence with the development of language. This is an example of the strong innatist hypothesis, and I investigated it in a historical study of the development of the verbal number sequence.

There is no a priori reason to exclude the verbal number sequence from the historical development of language because a system that specifies the semantic and syntactic properties of the number sequence surely must be more exact but simpler than the more general system alluded to by Chomsky.

That is, the rules that generate the verbal number sequence should not be somehow different from the rules that generate sentences.

In a historical development of the verbal number sequence, I would not expect a fully developed sequence to have emerged in one fell swoop in primitive cultures. However, if the strong innatist hypothesis is credible, it would be necessary to observe sequential organization as a basal intuition. Menninger claims just the opposite--two and three did not develop in such a way that they were elements of a sequential order.

Menninger (1969) distinguishes between two as dual and two as unity. The analysis of two as dual is a prenumerical analysis and, as such, is quite consistent with Brouwer's (1913) analysis of "two-oneness."

> This neo-intuitionism considers the falling apart of moments of life into qualitatively different parts, to be reunited only while remaining separated by time as the fundamental phenomenon of the human intellect, passing by abstracting from its emotional content into the fundamental phenomenon of mathematical thinking, the intuition of the bare two-oneness, the basal intuition of mathematics. (p. 85)

This "falling apart of moments of life into qualitatively different parts" is caught by Menninger (1969) as an awakening of consciousness--the isolation of self in an environment--"the I is opposed to and distinct from what is not I, the thou, the other" (p. 13). Both of these analyses of two as dual have been accounted for by von Glasersfeld (1981) in terms of an attentional mechanism. Von Glasersfeld's analysis has not yet been put into this context and has too often been passed over without comment. It should be understood as a model for a minimal mechanism that is necessary for moments of life to "break apart" or for the self to separate experience into "the I and the other."

In two as a unity, "we experience the very essence of number more intensely than in other numbers, that essence being to bind many together into one, to equate plurality and unity" (Menninger, 1969 p. 13). In the early developments, neither two as dual nor two as unity were separated from their sensory contents and, thus, were not the abstract two, a noun. Two was used as an adjective--two cows or two men. But two as unity cannot be a property of two cows because it cannot be a property of a single cow. So, it is not the result of an inductive inference and cannot be found in nature. Rather, it is the result of an operation of the mind that binds a dual together into a unity. The dual is a sensory-motor structure whose elements are qualitatively distinct from two as unity because, as pointed out by Brouwer (1913), two as unity requires an abstraction.

In the step to three, a new element appears in the dichotomy of I--You, namely, what lies beyond them, It. According to Menninger, the It is many.

> This statement, in which psychological, linguistic, and numerical elements come together, may perhaps roughly paraphrase early man's thinking about numbers. "One-two--many": a curious counting pattern, but it is mirrored in the grammatical number forms of the noun, singular--dual--plural." (pp. 16-17)

Menninger shows how the writing of the Egyptians perpetuated the early conceptual stage of the It as many. For example, water was symbolized as three waves. He takes the step to many as the decisive one in the development of the number sequence, because it introduced (but did not complete) infinite progression. It was a first step in the abstraction of the number sequence, an abstraction characterized by Menninger (1969, p. 7) as creating great difficulties for the human mind. Von Glasersfeld's (1981) model of the construction of plurality shows how plurality is a non-inductive construction and, along with the subsequent constructions that lead to the number sequence, provides insight into why the number sequence proved to be a difficult construction for the human mind.

This completes my discussion of Menninger's account of the historical development of the number sequence. I could find nothing in his study that convinced me of the necessity of viewing the number sequence as being innate. Of course, it is easy to say that one should have expected as much. But, if it had been possible to find three as the basal intuition, this would have been compatible with the strong innatist hypothesis. As explained by Menninger (1969), however, three required at least two previous abstractions, the first which transformed two as dual into two as unity and the second which introduced a new element into the contrast of I--You, namely, the It or many. The basal intuition of mathematics isolated by Brouwer is a construction, and it is fortunate to be able to appeal to a minimal attentional mechanism that could be used to account for this basal intuition as well as for the child's construction of object concepts (Steffe & von Glasersfeld, 1988).

Learning the Initial Number Sequence

I find it particularly striking that Menninger's historical analysis of the number sequence is compatible with an analysis of children's construction of the verbal number sequence (Steffe, 1988c). Five learning stages and the mechanisms of progress from one stage to another have been isolated. These stages are the perceptual counting scheme, the figurative counting scheme, the initial number sequence, the tacitly nested number sequence, and the explicitly nested number sequence. The counterexample to the learning paradox is taken from my explanation for how a child named Tyrone made progress from his figurative counting scheme to his initial number sequence.

A Figurative Counting Scheme

To illustrate the sense in which Tyrone's counting scheme was a figurative scheme, a protocol from an interview with him held on 15 October 1980 is provided. In the protocol, "T" is used to indicate his teacher's language and actions and "Ty" is used to indicate Tyrone's utterances and actions.

Protocol 1

T: (Places a card covered by two cloths in front of Tyrone) there are eight squares here (touches one of the cloths) and three here (touches the other cloth).
Ty: 1, 2, ..., 8 (looks at and synchronously touches the first cloth) 9, 10, 11 (looks at and synchronously touches the second cloth, his points of contact forming a linear pattern)--11.

There are several important features of Tyrone's counting behavior. His activity of looking at and touching the first cloth in distinct locations each time he performed a counting act indicates that he was aware of the items of the hidden collection, an awareness that was made possible by *re-presentation,* a recreation of an experience of an item without actual perceptual material. The recreated perceptual item is called a *figural unit item* (Steffe, von Glasersfeld, Richards, & Cobb, 1983). Figural unit items are not limited to visual imagery because perceptual material in all sensory channels can be re-presented.

As Tyrone counted over the cloth, we have to decide whether his number words and pointing acts referred to figural unit items or to *abstract unit items*. The latter are created using figural unit items by "running through" the figural unit items with a *unitizing* operation, stripping them of their sensory quality. Von Glasersfeld's (1981) attentional mechanism provides a model of the unitizing operation used in creating abstract unit items.

The same operation children use to create concepts of objects and their instances, the perceptual items of experience, is used to create abstract unit items. The operation does not change, but the material of application does. So, the important question is whether Tyrone was limited to creating and counting figural unit items or whether he could create and count abstract unit items as well. It is crucial for both theory and educational practice to be able to distinguish behaviorally between children whose counting scheme is figural and those whose counting scheme is numerical.

Counting abstract unit items, according to Menninger (1969) "constitutes *assigning* words to things" (p. 7). As productive activity, counting involves the vocal production of number words in a conventional sequence and their individual coordination to unit items of one kind or another. "In short, any distinguishable entity, tangible or intangible, identical or different, can be counted" (Menninger, 1969, p. 7). Menninger thus likens the terms of the number sequence to "small boxes" labeled 1, 2, 3, and so on, and the counting process as placing tangible items in the boxes, starting with the first box and proceeding to the next, and so on, until the items to be counted are exhausted. The last box filled indicates the numerosity of the items counted. Rather than use "boxes," I use "abstract unit items" and change "placing tangible items into boxes" to "re-presenting or reenacting the sensory-motor items pointed to by the records in the abstract unit items." The counting process can then be, for example, one of coordinating the production of a number word sequence with the reenactment of the motor acts pointed to by the records in a sequence of abstract unit items. Because it is possible for a number word to be recorded in an abstract unit item, uttering the number word would symbolize the abstract unit item as well as any other countable item so recorded.

Whatever the nature of the items were that Tyrone counted when counting over the first cloth, he coordinated the utterance of the number word sequence "1, 2, ..., 8" with repeatedly touching the cloth. These pointing acts served in isolating possible locations of hidden perceptual items and were the sensory items he created in the activity of counting. As such, they were substitutes for the countable items that were hidden and it is no exaggeration to say that Tyrone was counting his pointing acts. Still, the pointing acts may have symbolized abstract unit items. Although his failure to count on can be used as an argument against that interpretation, it cannot be taken as definitive, and we have to consider his continuation of counting past "eight" for a more solid argument.

Figurative and numerical patterns. Tyrone stopped counting over the second cloth when he recognized completing a linear spatial pattern for "three." Whether this pattern was a *figurative* or a *numerical* pattern turns on whether he took his records of pointing (there were no sensory traces) as material for the unitizing operation. To decide, I presented Tyrone with a collection of items again hidden by two cloths, one hiding seven and one hiding five squares.

Protocol 2

T: Tyrone, there are seven here (points to one cloth) and five here (points to the other).

Ty: (Touches the first cloth seven times while whispering) 1, 2, 3, 4, 5, 6, 7. (Tyrone's points of contact form no identifiable pattern. He continues touching the second cloth six times in a row while whispering) 8, 9, 10, 11, 12, 13.

Ty: (Realizes he doesn't know how many times he touched the second cloth, so he starts over without suggestion. In the midst of touching the second cloth, he loses track, so he starts over once again. This time he deliberately touches the second cloth five times in a row, while looking intently at his points of contact) 8, 9, 10, 11, 12 (looks up), 13, 14. Fourteen!

The way in which Tyrone uttered "13, 14" indicated that he wasn't sure whether he had just completed a pattern for "five." Nevertheless, he intentionally monitored how many times he touched the cloth. Starting over twice indicates that he tried to recognize a completed pattern for "five"; he tried to keep track of his counting acts.

The act of simply recognizing a row of five dots does not require an intentional monitoring of the activity that produces the pattern. When Tyrone was producing a pattern consisting of the records of his points of contact of his finger on a cloth, he had to establish what pattern he produced after each touch because there were no visible traces of his touches (it is not indicated in the protocol, but Tyrone was in deep concentration as he looked intently at the cloth). After he uttered "13 14," I asked Tyrone how many there were hidden under the second cloth. He said "five" and proceeded to count one more time.

Protocol 3

Ty: (Touches the first cloth and utters) 1, 2, 3, 4, 5, 6, 7, (continues touching the second cloth five times in a row, but this time stares into space) 8, 9, 10, 11--(looks at me) 12!

His conviction that he was to stop at "twelve," coupled with staring into space, again indicates intentional monitoring of counting activity. His efforts to organize his continuation of counting in Protocol 2 apparently led to his knowing when to stop counting in Protocol 3. But Tyrone clearly was yet to learn to count on. He had no concept of "seven" that he could re-present--he had just constructed one for "five"!

A re-presentation of a sensory-motor pattern is a recreation of the pattern without actual sensory-motor signals being available. Its product is a figurative (or re-presented) pattern. When Tyrone was counting over the second cloth trying to keep track of his pointing acts in Protocol 2, he had to create figurative patterns for the simple reason that there were no visual traces of counting. I have diagramed the situation in Figure 1 after Tyrone had counted "8, 9" and when it was his intention to continue to count until completing a pattern for "five," a pattern he was yet to associate with the number word.

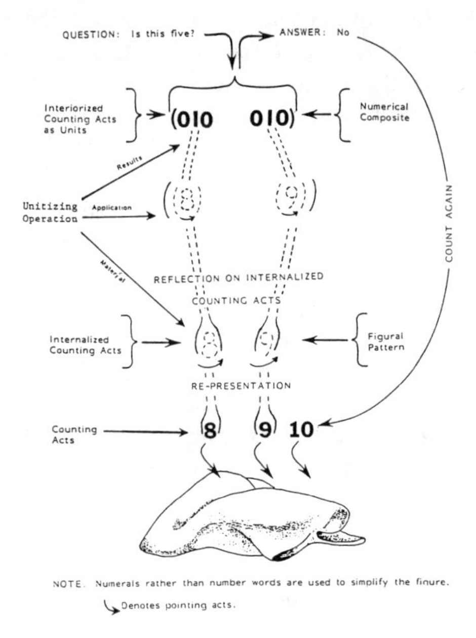

Figure 1: Tyrone's Engendering Accommodation

Re-presentation alone is insufficient to explain Tyrone's monitoring behavior, because three other children whose counting scheme was figurative created figurative patterns by re-presenting perceptual patterns and then used these figurative patterns to keep track of counting without *intentionally monitoring* their activity. It was extremely difficult to find situations that would foster the three children's becoming aware of their counting activity (Steffe, 1980b, p. 141). They never successfully created patterns to solve a problem as did Tyrone. To intentionally monitor counting activity there must be an explicit awareness of the patterns created in re-presentation while counting, which is to say the child must reflect on the patterns.

To explain how Tyrone became explicitly aware of the patterns he created to keep track of counting, my hypothesis is that he applied his unitizing operation to the pattern of re-presented counting acts. In doing so, my model is that he "ran through" the figurative pattern, applying the unitizing operation to its constituent elements. This is indicated by the dotted parentheses around the dotted numerals in Figure 1. This operation strips the figurative unitary items of their sensory-motor quality and creates a sequence of abstract unit items that contain the records of the sensory-motor material (abstract discreteness). I call this sequence of abstract unit items a numerical composite. But what it is called is not really essential to my current argument. The essential thing is that Tyrone created numerical patterns.

The concept that Tyrone finally created for "five" was, in my model of learning, a numerical pattern that contained the records of counting "8, 9, 10, 11, 12." So, in retrospect, my interpretation of the items that Tyrone's counting acts signified in Protocol 1 were figural unit items rather than abstract unit items, because there was no indication there that he reconstituted the involved counting acts as abstract unit items. His lack of a numerical concept for "seven" and "eight" was repeatedly corroborated in Protocols 1, 2, and 3 by his always starting to count with "one".

Monitoring as feedback. In Figure 1, there is a distinction between an internalized counting act and an interiorized counting act. An *internalized* counting act is a counting act that can be re-presented in the absence of actual sensory-motor material and an *interiorized* counting act is an abstract unit item that contains records of an internalized counting act. It too can be re-presented by re-presenting what the records point to. Application of the unitizing operation to the internalized counting acts "8, 9" disembeded them from the rest of the internalized counting scheme and recorded them at the level of interiorization. Tyrone, intending to continue on counting, could take the interiorized counting acts as belonging to the items he was counting, to the *countable items*. As internalized counting acts, they were *counted items*. The application of the unitizing operation thus made it possible for the counted items to feed back into the countable items and this is what I mean by monitoring in this case. Simply put, Tyrone reconstituted the counted items as countable items. His goal to continue on counting five times sustained this feedback system until his goal was reached.

A Verbal Number Sequence

By January 8, 1981, Tyrone had reorganized his counting scheme--he modified counting to fit his interpretation of the task in the following protocol.

Protocol 4

T: (Presents nine visible squares beside a cloth) Tyrone, you have 16 squares all together. Some of them are there (points to the visible squares). So, how many are here (pointing to the cloth)?
Ty: (Points to each visible square, subvocally uttering number words. He then rests his head on his hand while looking into space) eight!
T: Why did you say eight?
Ty: Because there's eight there and nine and eight make sixteen!
T: How do you know for sure?
Ty: (Sequentially puts up seven fingers while subvocally uttering number words) Seven!

In explanation of his answer "eight," Tyrone eliminated the redundant part of his counting activity and finally learned to count on! To do so, I agree with Bereiter that "nine" had to refer to a numerical concept. I will not go into what this concept consisted of now, because it is enough to say that learning to count on was based on this numerical concept whatever it consisted of. So, am I tumbling right into the learning paradox? In my reformulation of learning, the answer is NO.

Focusing on Tyrone's learning to count on when counting-up-to to find how many items were missing can blind one to the more important thing which Tyrone had learned. There was a change in Tyrone's use of his counting scheme that implies an underlying reorganization of the scheme. I had never observed him use counting-up-to prior to this observation nor had I observed him reorganize his counting scheme in any particular situation before the Christmas holidays--he could not solve my "missing items" situations. Rather, Tyrone reorganized his counting scheme over the Christmas holidays, when it was very unlikely that he solved anyone's missing items situation. I certainly had never modeled counting-up-to for him--his independently reorganized counting scheme was truly a novelty. Based on Tyrone's counting behavior on October 15, it would not be too much to say that his counting scheme had undergone a metamorphosis by January 8. How this metamorphosis is explained is essential in the reformulation of learning.

Autoregulation of monitoring. From Protocol 4, we see that Tyrone's number word sequence to at least "sixteen" was operative. To explain the metamorphosis from a figurative[3] to an operative[4] number word sequence, I hypothesize that a disturbance was created by part of his number word sequence being interiorized and the rest of it being internalized as diagramed in Figure 1. It existed on two levels. I further hypothesize that this perturbation was neutralized by autoregulation of the monitoring activity. The feedback system that was set in motion in experiential contexts continued until the process of interiorization of the internalized sequence of counting acts was completed. There are no assumptions made about how many times the unitizing operation was applied or how many adjacent number words might have served as material for a particular application. The only assumption that I make is that the unitizing operation was applied enough times to complete the interiorization process. The reorganization of his counting scheme appeared when Tyrone could use his interiorized number word sequence in assimilation of a missing items situation.

Accommodations and the Learning Paradox

The accommodation of Tyrone's figurative counting scheme that engendered the *metamorphic accommodation* was *monitoring a continuation of counting* and the metamorphic accommodation was *autoregulation of the process of interiorization*. The relevance of experience in the engendering accommodation is incontestable, because Tyrone created a numerical concept for "five" using "8, 9, 10, 11, and 12" as material. It is an example of what I call a *functional accommodation*--an accommodation of a scheme that occurs in the context of using the scheme.

The example of the learning paradox concerning counting given by Bereiter disappears in the face of Tyrone's engendering and metamorphic accommodations. The underlying reorganization of his counting scheme was a result of the operations of re-presentation and unitizing, operations sufficient to account for Tyrone's construction of his verbal number sequence, a conceptual system that was richer than his previous figurative counting scheme. It should be apparent that it was the *functioning* of re-presentation and the unitizing operation that produced the verbal number sequence. One only has to observe that they are the same operations of intelligence that yield children's concepts of the items of experience for refutation of the claim that re-presentation and the unitizing operation are somehow as complex as the verbal number sequence.

A More General Reformulation of Learning

Without engendering and metamorphic accommodations, it would be very implausible that children would ever develop meaning for number words like "ninety-seven." Invoking a process of inductive inference is insufficient because of the necessity to interiorize the internalized counting acts. In Tyrone's case, the interiorized material to which "five" initially pointed included a record of the lexical items "8, 9, 10, 11, 12" and there was just no possibility that he could somehow *infer* that such interiorized counting acts could be the meaning of, say, "sixteen." Moreover, although I have not documented it here, Tyrone became able to *insert* his numerical concept of number words at any point in his verbal number sequence (Steffe & Cobb, 1988, pp. 162 ff). For example, the material of "ten" could be "31, ..., 40" or any other segment of the verbal number sequence of numerosity ten (and this was not restricted to "ten"). What this means is that Tyrone used his verbal number sequence as material of his *uniting operation*, another engendering accommodation that cannot be explained by inductive inference.

We have seen above that both Menninger and Brouwer isolated the mental operation of binding many items together into one as the operation that creates composite units, an operation that was described by Joannes Caramuel in 1670.

> The intellect ... does not find numbers but makes them; it considers different things, each distinct in itself; and intentionally unites them in thought. (p. 44) Translated from Italian by E. von Glasersfeld.

The uniting operation has its origin in the attentional mechanism comprising the unitizing operation (Steffe, et al., 1983, p. 67). It is simply the unitizing operation applied to more than one figural or perceptual unit item. A number word usually comes to symbolize results of uniting operation--a composite unit--as well as an abstract unit item in a sequence of such items. The uniting operation is inextricably involved in the construction of numerical knowledge.

Accommodation and Learning

An engendering accommodation of a scheme leads to a modification of the scheme that is self-initiated, involves using conceptual elements or operations external as well as internal to the scheme, leads to further accommodations, and involves or leads to a structural reorganization of the scheme (Steffe, 1988b, p. 287). We see that Tyrone's engendering accommodation diagramed in Figure 1 was self-initiated and occurred as a result of goal-directed counting activity, involved using re-presentation and the unitizing operation, and led to a structural reorganization of his counting scheme--to his metamorphic accommodation.

A metamorphic accommodation of a scheme leads to a modification of the scheme that occurs independently but not in any particular application of the scheme. It is preceded by engendering accommodations (Steffe, 1988b, p. 287). In retrospect, Tyrone's engendering and metamorphic accommodations are but particular cases of the "reflecting" and "reflection" aspects of Piaget's (1980b) reflective abstraction.

> Logical-mathematical abstraction ... will be called "reflective" because it proceeds from the subject's actions and operations ... we have two interdependent but distinct processes: that of projection onto a higher plane of what is taken from the lower level, hence a "reflecting," and that of a "reflection" as a reorganization on the new plane. (p. 27)

In Tyrone's engendering accommodation of his figurative counting scheme, the "higher plane" consisted of the interiorized segment of his figurative number word sequence "8, 9, 10, 11, 12" and the "lower level" consisted of his internalized number word sequence. The operations that made these two levels distinguishable were the unitizing operation and re-presentation, respectively. The reorganization that took place "on the new plane" is explained by the metamorphic accommodation. Specifying engendering and metamorphic accommodations of specific mathematical schemes is crucial in formulating models of reflective abstraction and in developing a psychology of mathematics learning.

Abduction vs induction. In the context of scheme theory, engendering and metamorphic accommodations are reformulations of Peirce's notion of *abduction:* broadly speaking, abduction covers "all the operations by which theories and conceptions are engendered" (Fann, 1970, p. 8).

> In induction we generalize from a number of cases of which something is true and infer that the same thing is probably true of a whole class. But in abduction, we pass from the observation of certain facts to the supposition of a general principle to account for the facts. (Fann, p. 10)

In Fann's characterization of abduction, there was no mention made of whether the individual is aware of how the general principle that accounts for the facts might be arrived at or even if the individual is aware of the general principle. This ambiguity makes it important in the context of scheme theory to reformulate abduction as engendering and metamorphic accommodations. Tyrone certainly was not aware of how he counted on January 8 any more than was he aware of how he counted on October 25, nor was he aware of how he made progress. This is compatible with Piaget's (1980b) belief that "reflection comes into action through a set of still instrumental assimilations ... without any conceptual awareness of structures as such (this is to be found all through the history of mathematics)" (p. 28).

Piaget (1980b) formulated inductive inference in terms of empirical abstraction rather than reflective abstraction.

> Abstraction and generalization are obviously interdependent, each founded on the other. It results from this that only inductive generalization, proceeding from "some" to "all" by simple extension, will correspond to empirical abstraction, whereas constructive ... generalization in particular corresponds to reflective and reflected abstraction. (p. 28)

This is a reformulation of inductive inference as defined by Fodor because he did not specify the type of abstraction that is involved. Piaget (1980) defines empirical abstraction as "the kind that bears on physical objects external to the subject" (p. 27). It is crucial to note that the individual's conception of physical objects on which empirical abstraction bears are themselves the product of the unitizing operation. Consequently, the child's conception of physical objects as externalized, permanent objects that can have an existence in their own space and time independent of the child (Piaget, 1937; Inhelder & Piaget, 1957) is based on reflective abstraction (von Glasersfeld, 1981).

It should come as no surprise that Chomsky (1980a) doesn't know what reflective abstraction means.

> My uneasiness with "reflective abstraction" is ... that I do not know what the phrase means, to what processes it refers, or what are its principles. (p. 323)

This rejection of reflective abstraction and the concomitant acceptance of learning as inductive inference leads us straight into the learning paradox unless, of course, radical innatism is embraced to explain such phenomena as the metamorphosis of Tyrone's counting scheme. It seems more productive to embrace a weak innatist hypothesis in the sense explained by Piaget (1980b, p. 23) and to deny that learning is only inductive inference.

Procedural accommodation. It is not possible to characterize all mathematical learning in terms of engendering or metamorphic accommodations and Piaget's reformulation of inductive inference. For example, I have helped children learn to count on in a way that is compatible with the explanation of Klahr and Wallace. In these cases, it was always necessary to attribute numerical concepts to the children because of the rapidity with which they eliminated redundant procedures they had learned earlier, the most prominent being the use of perceptual finger patterns. These accommodations of the counting scheme were functional accommodations that were not engendering nor were they metamorphic. Even so, they are in the province of reflective abstraction.

I call the type of functional accommodation identified by Klahr and Wallace *procedural*. In a procedural accommodation there is a modification of the activity of the scheme or a novel way to view the results of the scheme, but with no immediate change in the assimilatory operations of the scheme. The type of procedural accommodation they identified arises as a result of using conceptual elements or operations internal to a scheme.

For example, given that a child has already constructed the verbal number sequence, if the child learns to count on by eliminating redundant counting, this would be a product of the operations of which the child is already capable. The child might use the results of counting as material of the uniting operation in a review of counting the elements of two collections. Say the child counted "1, 2, 3, 4, 5, 6"--"7, 8--9, 10--11" to find how many marbles in two sacks, one holding six and the other five marbles. If the child took the records of counting each nonvisible collection as material of the uniting operation, the child would create two composite units. These units could then be used as objects of reflection in a review of counting to establish connections between "six", "five", and "eleven." The procedural accommodation would be the application of the uniting operation to the records of counting, which would make curtailment of counting possible.

If the curtailment of counting was permanent--i.e., the next time the child assimilated what to an adult is an addition situation, operations of the modified scheme would be used in re-presentation but not enacted through actually counting (i.e., the child would count on)--this would be what Piaget called a constructive or assimilating generalization. It is important to note that what was abstracted to eliminate redundancy of counting had to do with the child's activity of counting.[5]

Like engendering accommodations, procedural accommodations are functional accommodations. In fact, some engendering accommodations are procedural (e.g., Tyrone's engendering accommodations) because they involve a modification of the activity or results of counting but lead to no immediate change in the assimilating operations of the scheme. But applying the unitizing operation to re-presented counting acts was novel in Tyrone's case, whereas in the example of eliminating a redundancy in counting[6], the application was not novel and was applied to interiorized records of counting. Although this may not lead to a change in the assimilating operations of the scheme, the child might count on in a similar situation. I would not call this induction in Piaget's reformulation because what was abstracted had to do with the child's activity of counting, not a property of some physical object in a collection of perceptual items. Hence, the abstraction is in the province of Piaget's reflective abstraction rather than empirical abstraction, and I would correspondingly place it in the realm of abduction rather than induction.

Retrospective accommodation. Some engendering accommodations involve a modification of the first part of a scheme prior to activity (cf. Steffe, 1988b, pp. 290-91) and in this sense are not procedural. They are examples of what I call retrospective accommodations: accommodations that are self-initiated and involve using conceptual elements constructed in an earlier application of the counting scheme (cf. Steffe, 1988b, p. 296). I have also isolated retrospective procedural accommodations that are not engendering (cf. Steffe, 1988a, pp. 133-34). My observations of retrospective accommodations, including engendering accommodations that are not procedural, involved abrupt changes either in the items to be counted or in the items being counted and were made by children who were yet to interiorize their counting scheme. Even with this restriction, I could not explain the changes on the basis of empirical abstraction and hence inductive inference.

It is important to note that I have not discussed deduction, the type of learning that seems to be erroneously considered as almost totally dominating mathematical learning. All that I do here is to note that reflected abstraction provides a possibility for its reformulation.

Final Comments

Mathematics educators traditionally have taken induction and deduction as models of learning and have essentially ignored schemes and their accommodations. Those who opt for this traditional view should carefully weigh the implications of the necessity to accept the radical innatist assumption that mathematics has to be conceived of as a mental entity that is programmed to unfold over time. Opting to accept the functioning of intelligence as being innate and reflective and empirical abstraction as a model of learning has potential for spawning creative work in mathematics education. But, rather than calling the "new" model of learning only "noninductive learning" like hyperbolic geometry is called "non-Euclidean geometry," following von Glasersfeld (1983), I call it "constructive learning" as well, a phrase that has its roots in Piaget's (1970a) genetic epistemology.

The creative work in constructive learning theory will emerge as models of reflective abstraction in the case of particular schemes. But reflective abstraction must be operationally defined in particular contexts with respect to particular schemes before it has any clear meaning. Percy Bridgman's (1934) well known position that "the meaning I ascribe to 'beautiful', for example, I find in the operations which I perform, or more simply, in what I do" (p. 104), has far-reaching implications for specifying operations like the unitizing operation that can be used in operationally defining reflective abstraction. Such a specification should be one of the primary intentions of researchers in mathematics education who work within constructivist learning theory.

Fodor (1980a), in summing up his argument, commented, "What I think it shows is really not so much an a priori argument for nativism as that there must be some notion of learning that is so incredibly different from the one we have imagined that we don't even know what it would be like as things now stand" (p. 269). Constructive learning is "incredibly different" from inductive learning and I can see no reason to declare a moratorium on the development of a theory of mathematical learning until a "man of genius" shows us the way, because Piaget already has. It is not that I have a bias or a prejudice that would lead to an uncritical acceptance of reflective abstraction--I just find it to be incredibly useful as a guiding heuristic in a search for insight into mathematical learning.

Notes

1. In a personal communication, Ernst von Glasersfeld pointed out that hypothesis formation can be a non-inductive activity. The question of the source of hypothesis formation cannot be satisfactorily answered without some kind of construction. If that is the case, a hypothesis that is confirmed becomes a piece of new knowledge, and it may be less or more complex than anything that was known before the event.
2. If Bereiter accepts Fodor's definition of inductive inference, he implicitly includes presuppositions that may well enable the child to construct a notion that fits the concept of cardinality in the given context. My agreement rests on a different understanding of inductive inference.
3. An internalized sequence of counting acts is a more or less permanent record of the kinesthetic, visual, and auditory aspects of past counting acts prior to their being reprocessed using the unitizing operation. Because the child uses his or her standard number word sequence in counting, an internalized sequence of counting acts is nothing more than an internalized number word sequence (von Glasersfeld, Steffe, & Richards, 1983, p. 26) whose lexical items signify countable items--a *figurative* number word sequence.
4. An interiorized sequence of counting acts is a sequence of abstract unit items that contain the records of the internalized sequence of counting acts that served as material in its construction. By a *verbal number sequence*, I mean a re-presentation of an interiorized sequence of counting acts using the auditory records--an *operative* number word sequence.

5. There are other possible explanations for the elimination of redundancy of counting. The unitizing operation is the essential element in all of these explanations.
6. The unitizing operation would be applied to serve in the achievement of the goal of finding, say, six and five more.

4 Abstraction, Re-Presentation, and Reflection: An Interpretation of Experience and Piaget's Approach

Ernst von Glasersfeld

> The understanding, like the eye, whilst it makes us see and perceive all other things, takes no notice of itself; and it requires art and pains to set it at a distance and make it its own object.
>
> John Locke (1690)[1]

> As adults we are constantly deceiving ourselves in regard to the nature and genesis of our mental experiences.
>
> John Dewey (1895)[2]

One of the remarkable features of the Behaviorist era in American psychology is that so many leaders and followers of that creed could claim to be Empiricists, cite John Locke as their forefather, and get away with it. Had they read the first chapter of "Book II"[3] of his major work, *An Essay Concerning Human Understanding*, they would have found, among many others, the following enlightening statements. Paragraph 2 has the heading:

All Ideas Come From Sensation or Reflection.

Paragraph 4 has the heading *The Operations of Our Minds,* and it is there that Locke explains what he means by "reflection":

> This source of ideas every man has wholly in himself; and though it be not sense, as having nothing to do with external objects, yet it is very like it, and might properly enough be called *internal sense*. But as I call the other Sensation, so I call this *Reflection*, the ideas it affords being such only as

> the mind gets by reflecting on its own operations within itself.

In our century, it was Jean Piaget who vigorously defended and expanded the notion of reflection. He lost no opportunity to distance himself from empiricists who denied the mind and its operations and wanted to reduce all knowing to a passive reception of objective "sense data." Yet, he should not have found it difficult to agree with Locke's division of ideas because it is not too different from his own division between *figurative* and *operative knowledge*. Similarly, I feel, Locke would have had a certain respect for Piaget's effort to set understanding at a distance and to make it the object of investigation. And both men, I have no doubt, would have agreed with Dewey about the risk of deceiving oneself by taking mental experiences as given. It is therefore with caution that I shall proceed to discuss, in the pages that follow, first my own view of reflection, abstraction, re-presentation, and the use of symbols, and then a tentative interpretation of Piaget's position. If, at times, I may sound assertive, I would beg the reader to keep in mind that I am fully aware of the fact that I am merely offering conjectures--but they are conjectures which I have found useful in constructing a model of mental operations.

Reflection

If someone, having just eaten an apple, takes a bite out of a second one, and is asked which of the two tasted sweeter, we should not be surprised that the person could give an answer. Indeed, we would take it for granted that under these circumstances any normal person could make a relevant judgment.

We cannot observe how such a judgment is made. But we *can* hypothesize some of the steps that seem necessary to make it. The sensations that accompanied the eating of the first apple would have to be remembered, at least until the question is heard.[4] Then they would have to be re-presented and compared (in regard to whatever the person called "sweetness") with the sensations accompanying the later bite from the second apple. This re-presenting and comparing is a way of operating that is different from the processes of sensation that supplied the material for the comparison. Reflecting upon experiences is clearly not the same as having an experience.

A hundred years after Locke, Wilhelm von Humboldt wrote down a few aphorisms which, posthumously, his editors put under the heading "About Thinking and Speaking." The first three aphorisms deal with reflection:

1. The essence of thinking consists in reflecting, i.e., in distinguishing what thinks from what is being thought.

2. In order to reflect, the mind must stand still for a moment in its progressive activity, must grasp as a unit what was just presented, and thus posit it as object against itself.
3. The mind compares the units, of which several can be created in that way, and separates and connects them according to its needs (1907, Vol. 7, part 2, p. 581)[5].

I know of no better description of the mysterious capability that allows us to step out of the stream of direct experience, to re-present a chunk of it, and to look at it as *though* it were direct experience, while remaining aware of the fact that it is not. I call it mysterious because, although we can all do it as easily as flipping a switch, we have not even the beginnings of a model (least of all an "information processing" model) that would suggest *how* it might be achieved.

"To grasp as a unit what was just presented" is to cut it out of the continuous experiental flow. In the literal sense of the term, this is a kind of *abstraction*--namely the simplest kind. Focused attention picks a chunk of experience, isolates it from what came before and from what follows, and treats it as a closed entity. For the mind, then, "to posit it as object against itself," is to *re-present* it.

In the next two sections, I want to deal with abstraction and re-presentation one after the other.

Abstraction

As von Humboldt stated in his third aphorism, chunks of experience, once isolated, can be compared, separated, and connected. This makes possible further steps of *abstraction*, among them the kind that Piaget and many others have called "generalizing abstraction." Because it seems crucial in all forms of naming and categorization, it has been discussed for a long time. To clarify the core of the notion, I once more return to Locke, because he produced a very simple and widely accepted description of the process:

> This is called *Abstraction*, whereby ideas taken from particular beings become general representations of all the same kind; and their names general names, applicable to whatever exists conformable to such abstract ideas. (Book II, Ch. X, §9)

Locke's use of the words "being" and "exist" in this context caused Berkeley, who had a very different view of "existence," to voice a sarcastic objection against his predecessor.

> Whether others have this wonderful faculty of abstracting their ideas, they best can tell; for myself, I find indeed I have a faculty of imagining, or representing to myself, the ideas of those particular things I have perceived, and of variously compounding them. I can imagine a man with two heads, or the upper parts of a man joined to the body of a horse, I can consider the hand, the eye, the nose, each by itself abstracted or separated from the rest of the body. But then whatever hand or eye I imagine, it must have some particular shape and colour. (1710, Introd., §10)

This passage is interesting for two reasons. Berkeley claims, much as did von Humboldt, that we are able to re-present to ourselves particular experiential items and that we are also able to segment them and to recombine the parts at will. Then however, he goes on to claim that whatever we re-present to ourselves must have *the character of a particular*--from which he concludes that we cannot have *general* ideas.

Both these claims concern *re-presentation* and are, I believe, perfectly valid. But what follows from them is that we are unable to *re-present* general ideas, not that we cannot *have* them. Berkeley, it seems, unwittingly trapped himself into this position about abstraction. At the beginning of his *Treatise*, he says among other things:

> Thus, for example, a certain colour, taste, smell, figure and consistence having been observed to go together, are accounted one distinct thing, signified by the name *apple*; other collections of ideas constitute a stone, a tree, a book, and the like sensible things. (1710, §1)

Berkeley, of course, was aware of the fact that he would apply the name "apple" not only to one unique "thing," but to countless others that fitted his description in terms of "colour, taste, smell, figure, and consistence," but it seems that he took this ability quite simply for granted. Had he analysed it the way he analysed so many other conceptual operations, he might have changed his view about abstraction. I hope to make this clear with the help of an example.

A child growing up in a region where apples are red would necessarily and quite correctly associate the idea of redness with the name "apple." A distant relative arriving from another part of the country, bringing a basket of yellow apples, would cause a major perturbation for a child, who might want to insist that yellow things should not be called "apples." However, the social pressure of the family's usage of the word will soon force the child to accept the fact that the things people call "apple" come in different colors. The child might then be *told* that apples can also be green, which would enable the child to *recognize* such a particular green thing as an apple the first time it is brought to the house.

Berkeley, I would say, was quite right when he maintained that every time we imagine an apple, it has to have a specific color, but he was wrong to claim that we could, therefore, not have a *general idea* in our heads that allows us to recognize as apples items that differ in some respect, but nevertheless belong to that class.

Hence I suggest that, *pace* Berkeley, we are quite able to abstract *general ideas* from experience and that we do this by substituting a kind of place-holder or *variable* for some of the properties in the sensory complex we have abstracted from our experiences of particular things. I see no reason why one should not call the resulting cognitive structure a *concept*. Such a structure is more specific with regard to some properties and less specific with regard to others, and it is precisely because of this relative indeterminacy that it enables us to *recognize* items that we have never seen before as exemplars of a familiar kind.

In short, in order to recognize several particular experiential items, in spite of differences they may manifest, as belonging to the same kind, we must have a concept that is flexible enough to allow for a certain variability. That is, instead of specific particulars, it must contain variables. Yet it is clear that, in order to "imagine" for instance an apple, we have to decide what color it is to be, because we cannot possibly visualize it red *or* green *or* yellow *at one and the same time*. Berkeley, therefore, was right when he observed that whenever we *re-present* a concept to ourselves, we find that it is a particular thing and not a general idea. What he did not realize was that the abstraction necessary to recognize things of a kind does not automatically turn into an image that can be re-presented.

The situation, however, is somewhat complicated by our ability to use *symbols*, but before considering this, I want to deal with re-presentation.

Re-Presentation

No act of mental re-presentation, which in this context of conceptual analysis means neither less nor more than the re-generation of a prior experience, would be possible if the original generation of the experience had not left some mark to guide its reconstruction. In this requirement, re-presentation is similar to recognition. Both often work hand in hand, e.g., when one recognizes a Volkswagen, though one can see only part of its back but is nevertheless able to visualize the whole. The ability to recognize a thing in one's perceptual field, however, does not necessarily bring with it the ability to re-present it spontaneously. We have all had occasion to notice this. Our experiential world contains many things which, although we recognize them when we see them, are not available to us when we want to visualize them. There are, for instance, people whom we would recognize as acquaintances when we meet them, but were we asked to describe them when they are not in our visual field, we would be unable to recall an adequate image of their appearance.

The fact that recognition developmentally precedes the ability to re-present an experiential item spontaneously, has been observed in many areas. It is probably best known and documented as the difference between what linguists call "passive" and "active" vocabulary. The difference is conspicuous in second-language learners, but it is noticeable also in anyone's first language: a good many words one knows when one hears or reads them are not available when one is speaking or writing.

This lag suggests that having abstracted a concept that may serve to recognize and categorize a perceptual item is not sufficient to re-present the item to oneself in its absence. Piaget has always maintained that all forms of imaging and re-presenting are, in fact, acts of internalized imitation[6] (Piaget, 1945). In this context, the metaphor of "program" may be useful. A program is the fixed itinerary of an activity that can guide and govern the sequence of its re-enactment. But there are two points to be stressed. First, a program may *specify* the material on which to act, but it does not supply the material; second, a program may *specify* what acts are to be performed, but it supplies neither the acting agent nor the action.

The first of these limitations, I suggest, may account for the fact that to recognize an experiential item requires less effort than to re-present it spontaneously. This would be so, because in re-presentation not only a program of composition is needed, but also the specific sensory components, which must be expressly generated. In recognition, the perceiver merely has to isolate the sensory elements in the sensory manifold. As Berkeley (1710) observed, sensory elements are "not creatures of the will" (§ 29). Because there are always vastly more sensory elements than the perceiving agent can attend to and use,[7] recognition requires the attentional selecting, grouping, and coordinating of sensory material that fits the composition program of the item to be recognized. In re-presentation, on the other hand, some substitute for the sensory raw material must be generated. (As the example of the Volkswagen indicates, the re-generation of sensory material is much easier when parts of it are supplied by perception, a fact that was well known to the proponents of Gestalt psychology.)

A difference analogous to that between acting on actually present perceptual material, as opposed to acting on material that must itself be generated, arises from the second limiting feature I mentioned. With regard to the need for an acting agent, a program is similar to a map. If someone draws a simple map to show you how to get to his house, he essentially indicates a potential path from a place you are presumed to know to the unknown location. The drawing of the path is a *graphic* representation of the turns that have to be made to accomplish that itinerary, but it does not and could not show what it is *to move* and what it is *to turn right or left*. Any user of the map must supply the motion and the changes of direction with the focus of visual attention while reading the map. Only if one manages to abstract this sequence of motions from the reading activity, can one transform it into physical movement through the mapped region. (Note that this abstracting and transforming is by no means an easy task for those unaccustomed to map reading.)

A program, however, differs from a map in that it explicitly provides instructions about actions and implicitly indicates changes of location through the conventional sequence in which the instructions must be read. (But in the program, too, it is the user's focus of attention, while reading or implementing the program, that supplies the progressive motion.) Also, unlike a map, a program may contain embedded "subroutines" for walking and turning (i.e., instructions *how* to act), but no matter how detailed these subroutines might be, they can contain only instructions to act, not the actions themselves. In other words, irrespective of how minutely a program's instructions have decomposed an activity, they remain static until some agent implements them and adds the dynamics.

In carrying out a program in an experiental situation, just as in following a map through an actual landscape, the sensory material in the agent's perceptual field can supply cues as to the action required at a given point of the procedure. In the re-presentational mode, however, attention cannot focus on actual perceptual material and pick from it cues about what to do next, because the sensory material itself has to be re-presented. A re-presentation--at least when it is a spontaneous one--is wholly self-generated (which is one reason why it is usually easier to find one's way through a landscape than to draw a reliable map of it).

The increase of difficulty and the concomitant increase of effort involved in the production of conceptual structures when the required sensory material is not available in the present perceptual field shows itself in all forms of re-presentation and especially in the re-enactment of abstracted programs of action. Any re-presentation, be it of an experiential "thing" or of a program of actions or operations, requires *some* sensory material for its execution. That basic condition, I believe, is what confirmed Berkeley in his argument against the "existence" of abstracted general ideas, for it is indeed the case that every time we re-present to ourselves such a general idea, it turns into a particular one because its implementation requires the *kind* of material from which it was abstracted.

This last condition could be reformulated by saying that there has to be some isomorphism between the present construct and what it is intended to *re*construct. Clearly, this isomorphism does not concern a "thing-in-itself" but precisely those aspects one wants to or happens to focus on. As Silvio Ceccato remarked thirty years ago, what we visualize of objects in dreams is no more than is required by the context of the dream.[8]

More importantly, this selective isomorphism is the basis of *graphic* and *schematic representations*. They tend to supply such perceptual material as is required to bring forth in the perceiver the particular ways of operating that the maker of the graphic or schematic is aiming at. In this sense they are *didactic* because they can help to focus the naive perceiver's attention on the particular operations that are deemed desirable. Hence, as I have suggested elsewhere (cf. von Glasersfeld, 1987), they can be divided into *iconic* and *symbolic* representations, but neither kind should be confused with the *mental* re-presentation I am discussing here.

The Power of Symbols

Re-presentations can be activated by many things. Any element in the present stream of experience may bring forth the re-presentation of a past situation, state, activity, or other construct. This experiential fact was called *association* by Hume and used by Freud for his analyses of neuroses. The ability to associate is systematically exploited by language. To possess a word is to have associated it with a representation of which one believes that it is similar to the re-presentations the word brings forth in other users of the language. (Only naive linguists claim that these re-presentations are *shared*, in the sense that they are the same for all users of the given word.) In my terminology, a word is used as a *symbol*,[9] only when it brings forth in the user an abstracted generalized re-presentation, not merely a response to a particular situation (cf. von Glasersfeld, 1974).

Several things, therefore, are indispensable for a word to function as a symbol. (1) The phonemes that compose the word in speech, or the graphic marks that constitute it in writing, must be recognized as that particular item of one's vocabulary. This ability to recognize, as I suggested earlier, is preliminary to the ability to re-present and produce the word spontaneously. (2) The word/symbol must be associated with a conceptual structure that was abstracted from experience and, at least to some extent, generalized. Here, again, the ability to recognize (i.e., to build up the conceptual structure from available perceptual material) precedes the ability to re-present the structure to oneself spontaneously.

Once a word has become operative as a symbol and calls forth the associated *meaning* as re-presentational chunks of experience that have been isolated (abstracted) and to some extent *generalized*, its power can be further expanded. By this I mean that, as particular users of the word become more proficient, they no longer need to actually *produce* the associated conceptual structures as a completely implemented re-presentation, but can simply register the occurrence of the word as a kind of "pointer" to be followed if needed at a later moment. I see this as analogous to the capability of recognizing objects on the basis of a *partial* perceptual construction. In the context of symbolic activities, this capability is both subtle and important. An example may help to clarify what I am trying to say.

If, in someone's account of a European journey, you read or hear the name "Paris," you may register it as a pointer to a variety of experiential "referents" with which you happen to have associated it--e.g., a particular point on the map of Europe, your first glimpse of the Eiffel Tower, the Mona Lisa in the Louvre--but if the account of the journey immediately moves to London, you would be unlikely to implement fully any one of them as an actual re-presentation. At any subsequent moment, however, if the context or the conversation required it, you could return to the mention of "Paris" and develop one of the associated re-presentations.

I have chosen to call this function of symbols "pointing" because it seemed best to suggest that words/symbols acquire the power to open or activate pathways to specific re-presentations without, however, obliging the proficient symbol user to produce the re-presentations there and then.

This function, incidentally, constitutes one of the central elements of our theory of children's acquisition of the concept of number (Steffe et al., 1983). In this theory, the first manifestation of an *abstract* number concept is a demonstration that the subject knows, *without carrying out a count*, that a number word implies or points to the sequential one-to-one coordination of all the terms of the standard number word sequence, from "one" up to the given word, to *some* countable items. Indeed, we believe that this is the reason why, as adults, we may assert that we know what, say, the numeral (symbol) "381,517" *means*, in spite of the fact that we are unlikely to be able to re-present to ourselves an associated particular chunk of experience; we know what it means, because it points to a familiar counting procedure or other mathematical method of arriving at the point of the number line which it indicates.

In mathematics this form of symbolic implication is so common that it usually goes unnoticed. For instance, when one is told that the side of a regular pentagon is equal to half the radius of the circumscribed circle multiplied by $\sqrt{10 - 2\sqrt{5}}$, one does not have to draw the square roots to *understand* the statement--provided one knows the operations the symbols "point to." The potential ability is sufficient; one does not have to carry out the indicated operations. Because it is so often taken for granted that mathematical expressions can be understood without carrying out the operations they symbolize, formalist mathematicians are sometimes carried away and declare that the manipulation of symbols constitutes mathematics.

Piaget's Theory of Abstraction

Few, if any, thinkers in this century have used the notion of abstraction as often and insistently as did Piaget. Indeed, in his view "All new knowledge presupposes an abstraction,..." (Piaget, 1974a, p. 89). But not all abstraction is the same. Piaget distinguished two main kinds, "empirical" and "reflective," and then subdivided the second. He has frequently explained the primary difference in seemingly simple terms, for example:

> Empirical abstractions concern observables and reflective abstractions concern coordinations. (Piaget, et al., 1977, Vol. II, p. 319)
>
> One can thus distinguish two kinds of abstraction according to their exogenous or endogenous sources... . (Piaget, 1974a, p. 81)

Anyone who has entered into the spirit of *Genetic Epistemology* will realize that the simplicity of these statements is deceptive. The expressions "observables" and "exogenous" are liable to be interpreted in a realist sense, as aspects or elements of an external reality. Given Piaget's theory of knowledge, however, this is not how they were intended. In fact, the quoted passages are followed by quite appropriate warnings. After the first, Piaget explains that no characteristic is in itself observable. Even in physics, he says, the measured magnitudes (mass, force, acceleration, etc.) are themselves constructed and are therefore results of inferences deriving from preceding abstractions (Piaget et al., 1977, Vol. II, p. 319). In the case of the second quotation, he adds a little later: "there can be no exogenous knowledge except that which is grasped as content, by way of forms which are endogenous in origin". (Piaget, 1974a, p. 91) This is not an immediately transparent formulation. As so often in Piaget's writings, one has to look elsewhere in his work for enlightenment.

Form and Content

The distinction between form and content has a history as long as Western philosophy, and the terms have been used in many different ways. Piaget's use of the distinction is complicated by the fact that he links it with his use of "observables" (content) and "coordinations" (forms). "The functions of form and content are relative, since every form becomes content for another that comprises it" (Piaget et al., 1977, Vol. II, p. 319). This will make sense only if one recalls that, for Piaget, percepts, observables, and any knowledge of objects are all the result of a subject's action and *not* externally caused effects registered by a passive receiver. In his theory, to perceive, to remember, to re-present, and to coordinate are all dynamic in the sense that they are activities carried out by a subject that operates on internally available material and produces certain results.

A term such as "exogenous," therefore, must not be interpreted as referring to a physical outside relative to a physical organism, but rather as referring to something that is external *relative to the process* in which it becomes involved.

Observation and re-presentation have two things in common: (1) they operate on items which, relative to the process at hand, are considered *given*. The present process takes them as elements and coordinates them as "content" into a new "form" or "structure"; and (2) the resulting new products can be taken as initial "givens" by a future process of structuring, relative to which they then become "content." Thus, once a process is achieved, its results may be considered "observables" or "exogenous" relative to a subsequent process of coordination on a higher level of analysis.

As Piaget saw, this might seem to lead to an infinite regress (Piaget et al., 1977, Vol. II, p. 306), but he put forth at least two arguments to counter this notion. One of them emerges from his conception of scientific analysis. Very early in his career, he saw this analysis as a cyclical program in which certain elements abstracted by one branch of science become the "givens" for coordination and abstraction in another. In an early paper (Piaget, 1929) and almost forty years later in his "classification of the sciences" (Piaget, 1967), he formulated this mutual interdependence of the scientific disciplines as a circle: biology--psychology--mathematics--physics, and looping back to biology. Hence, from his perspective, there is no linear progression without end, but simply development of method and concepts in one discipline leading to novel conceptualization and coordination in another. The recent impact of the physics of molecules and particles on the conceptual framework of biology would seem a good example.

Scheme Theory

The second reason against an infinite regress of abstractions is grounded in the developmental basis of Genetic Epistemology and is directly relevant here. The child's cognitive career has an unquestionable beginning, a first stage during which the infant assimilates, or tries to assimilate, all experience to such fixed action patterns (reflexes) as it has at the start (Piaget, 1975; p. 180). Except for their initial fixedness, these action patterns function like the *schemes* which the child a little later begins to coordinate on the basis of experience.

Schemes are composed of three elements: (1) an initial experiental item or configuration (functionally linked to what the observer would categorize as "trigger" or "stimulus"), (2) an activity the subject has associated with it, and (3) a subsequent experience associated with the activity as its outcome or result. Schemes thus govern the subject's segmentation and differentiation of experience.

When a novel item ("novel" in the *observer's* judgement) is assimilated to the initiating element of a scheme, it triggers the associated activity. If the activity leads to the expected result, the acting subject in no way differentiates the item from those that functioned like it in the past. But if, for one reason or another, the activity does *not* lead to the expected result, this generates a perturbation, which could be described as either disappointment or pleasant surprise. In either case, the perturbation may focus the subject's attention on the configuration that triggered the activity this time, and it may then be discriminated from those past experiences where the activity functioned in the expected manner (cf. Piaget, 1974b, p. 264). If the failure of the scheme and the ensuing discrimination of the novel item or situation leads to the tightening of the criteria of assimilation that determine what can and what cannot be taken as a trigger for the particular scheme, this would constitute an *accommodation* of the initiating conceptual structure.[10] Similarly, if the outcome is a pleasant surprise, this, too, may lead to an accommodation, in the sense that a *new* scheme will subsequently be triggered by the newly isolated experience.

In infancy, during the child's first two years, i.e., in the sensory-motor period, all this is assumed to take place without awareness and conscious reflection. Yet, the fact that three- or four-month-old infants *assimilate* items (which, to an observer, are not all the same) as triggers of a particular scheme is sometimes described as the ability to *generalize*. Indeed, this is what animal psychologists working with rats or monkeys call "stimulus generalization." With children at a later stage, however, when reflection has begun to operate, the discriminating of experiential items that do function in a given scheme from others that do not, constitutes a mechanism that functions as the source of *empirical abstractions* that are recognized as such by the acting subject. This, obviously, raises the question when and how the acting subject's awareness is involved.

One reason why that question seems quite urgent is that the word "reflection," ever since Locke introduced it into the human sciences, has implied a conscious mind that does the reflecting. A second reason is that in many places where Piaget draws the distinction between the "figurative" and the "operative," this tends to reinforce the notion that the operative (what Locke described as "the ideas the mind gets by reflecting on its own operations") requires consciousness. As a consequence, it would be desirable to unravel when, in Piaget's theory of the cognitive development, the capability of conscious reflection arises.

Piaget himself, as I have said elsewhere (von Glasersfeld, 1982), rarely makes explicit whether, in a given passage, he is interpreting what he is gathering from his observations (observer's point of view), or whether he is conjecturing an autonomous view from the observed subject's perspective (cf. Vuyk, 1981, Vol II). This difference seems crucial in building a model of mental operations and, therefore, to an understanding of his theory of *abstraction* and, especially, *reflective abstraction*. I shall return to this question of consciousness after the next section, where I try to lay out the kinds of abstraction Piaget has distinguished.

Four kinds of Abstraction

The process Locke characterized by saying, "whereby ideas taken from particular beings become general representations of all the same kind," falls under Piaget's term *empirical abstraction*. To isolate certain sensory properties of an experience and to maintain them as repeatable combinations, i.e., isolating what is needed to recognize further instantiations of, say, apples, undoubtedly constitutes an empirical abstraction. But, as I suggested earlier, to have composed a concept that can serve to recognize (assimilate) items as suitable triggers of a particular scheme does not automatically bring with it the ability to visualize such items spontaneously as re-presentations. Piaget makes an analogous point--incidentally, one of the few places where he mentions an empiricist connection:

> But it is one thing to extract a character, x, from a set of objects and to classify them together on this basis alone, a process which we shall refer to as "simple" abstraction and generalization (and which is invoked by classical empiricism), and quite another to recognise x in an object and to make use of it as an element of a different (non-perceptual) structure, a procedure which we shall refer to as "constructive" abstraction and generalisation. (Piaget, 1969, p. 317)

The capability of spontaneous re-presentation (which is "non-perceptual," too) develops in parallel with the acquisition of language and may lead to an initial, albeit limited form of awareness. Children at the age of three or four years are not incapable of producing *some* pertinent answer when they are asked what a familiar object is like or not like, even when the object is *not* in sight at the moment. This suggests that they are able not only to call forth an empirically abstracted re-presentation but also to review it quite deliberately.

The notion of empirical abstraction covers a wider range of experience for Piaget than is envisioned in the passage I quoted from Locke. What Locke called "particular beings" were for him "ideas" supplied by the five senses. Because, in Piaget's view, visual and tactual perception involve motion, it is not surprising that the internal sensations caused by the agent's own motion (kinesthesis) belong to the "figurative" and are, therefore for him, raw material for empirical abstractions in the form of motor patterns.[11]

That such abstracted motor patterns reach the level where they can be re-presented, you can check for yourself. Anyone who has some proficiency in activities such as running down stairs, serving in tennis, swinging for a drive in golf, or skiing down a slope has no difficulty in re-presenting the involved movements without stirring a muscle. An interesting aspect in such "dry reruns" of abstracted motor experiences is that they don't require specific staircases, balls, or slopes. I mention this because it seems to me to be a clear demonstration of deliberate and therefore conscious re-presentation of something that needed no consciousness for its abstraction from actual experience. This difference is important also in Piaget's subdivision of *reflective abstractions* to which we turn now.

From empirical abstractions, whose raw material is sensory-motor experience, Piaget, as I said earlier, distinguished three types of *reflective abstraction*. Unfortunately, the French labels Piaget chose for them are such that they are inevitably confused by literal translation into English.

The first "reflective" type derives from a process Piaget calls *réfléchissement*, a word that is used in optics when something is being reflected, as for instance the sun's rays on the face of the moon. In his theory of cognition, this term is used to indicate that an *activity* or *mental operation* (not a static combination of sensory elements) developed on one level is abstracted from that level of operating and applied to a higher one, where Piaget then considers it to be a réfléchissement (Moessinger & Poulin-Dubois, 1981, have translated this as "projection", which captures something of the original sense). But Piaget stresses that a second characteristic is required:

> Reflective abstraction always involves two inseparable features: a "réfléchissement" in the sense of the projection of something borrowed from a preceding level onto a higher one, and a "réflexion" in the sense of a (more or less conscious) cognitive reconstruction or reorganization of what has been transferred. (Piaget, 1975, p. 41)

At the beginning of the first of his two volumes on reflective abstraction (Piaget et al. 1977), the two features are again mentioned:

> Reflective abstraction, with its two components of "réfléchissement" and "réflexion," can be observed at all stages: from the sensory-motor levels on, the infant is able,

> in order to solve a new problem, to borrow certain coordinations from already constructed structures and to reorganize them in function of new givens. We do not know, in these cases whether the subject becomes aware of any part of this. (Piaget et al., 1977, Vol. I, p. 6)

In the same passage he immediately goes on to describe the second type of reflective abstraction:

> In contrast, at the later stages, when reflection is the work of thought, one must also distinguish thought as a process of construction and thought as a process of retroactive thematization. The latter becomes a reflecting on reflection; and in this case we shall speak of "*abstraction réfléchie*" (reflected abstraction) or *pensée réflexive* (reflective thought).

Since the present participle of the verb *réflechir*, from which both the nouns *réfléchissement* and *réflexion* are formed, is *réfléchissante*, Piaget used "*abstraction réfléchissante*" as a generic term for both types. It is therefore not surprising that in most English translations the distinction was lost when the expression "reflective abstraction" was introduced as the standard term.

The situation is further confounded by the fact that Piaget distinguished a third type of reflective abstraction which he called "pseudo-empirical." When children are able to re-present certain things to themselves but are not yet fully on the level of concrete operations,

> it happens that the subjects, by leaning constantly on their perceivable results, can carry out certain constructions which, later on, become purely deductive (e.g., using an abacus or the like for the first numerical operations). In this case we shall speak of "pseudo-empirical abstractions" because, in spite of the fact that these results are read off material objects as though they were empirical abstractions, the perceived properties are actually introduced into these objects by the subject's activities. (Piaget, et al., 1977, Vol. I, p. 6)

To recapitulate, Piaget distinguishes four kinds of abstraction. One is called "empirical" because it abstracts sensory-motor properties from experiental situations. Of the three "reflective" ones, the first projects and reorganizes on another level a coordination or pattern of the subject's own activities or operations. The next is similar in that it also involves patterns of activities or operations, but it includes the subject's awareness of what has been abstracted and is therefore called "reflected abstraction." The last is called "pseudo-empirical" because, like empirical abstractions, it can take place only if suitable sensory-motor material is available.

The Question of Awareness

One of the two main results of the research carried out by Piaget and his collaborators on the attainment of awareness, he summarized as follows in *La prise de conscience*[12]:

> ...action by itself constitutes an autonomous knowledge of considerable power, for while it is only "know-how" and not knowledge that is conscious of itself in the sense of conceptualized understanding, it nevertheless constitutes the source of the latter, because the attainment of consciousness nearly always lags quite noticeably behind this initial knowledge which is remarkably efficacious even though it does not know itself. (Piaget, 1974*b*, p. 275)

The fact that conscious conceptualized knowledge of a given situation developmentally lags behind the knowledge of how to act in the situation is commonplace on the sensory-motor level. In my view, as I mentioned earlier, this is analogous to the temporal lag of the ability to re-present a given item relative to the ability to recognize it. But the ability spontaneously to re-present to oneself a sensory-motor image of, say, an apple still falls short of what Piaget in the previous passage called "conceptualized understanding". This would involve awareness of the characteristics inherent in the *concept* of apple or whatever one is re-presenting to oneself, and this kind of awareness constitutes a higher level of mental functioning.

This further step requires a good deal more of what Locke called the mind's "art and pains to set (something) at a distance and make it its own object." A familiar motor pattern is once more a good example: we may be well able to re-present to ourselves a tennis stroke or a golf swing, but few, if any, would claim to have a "conceptualized understanding" of the sequence of elementary motor acts that are involved in such an abstraction of a delicately coordinated activity. Yet it is clear that, insofar as such understanding is possible, it can be built up only as a "retroactive thematization," that is, *after* the whole pattern has been empirically abstracted from the experience of enacting it.

In Piaget's theory, the situation is similar in the first type of *reflective* abstraction: he maintains that it, too, may or may not involve the subject's awareness.

> Throughout history, thinkers have used thought structures without having grasped them consciously. A classic example: Aristotle used the logic of relations, yet ignored it entirely in the construction of his own logic. (Piaget & Garcia, 1983, p. 37)

In other words, one can be quite aware of what one is cognitively operating on, without being aware of the operations one is carrying out.

As for the second type, "reflective thought" or "reflected abstraction," it is the only one about which Piaget makes an explicit statement concerning awareness:

> Finally, we call the result of a reflective abstraction "reflected" abstraction, once it has become conscious, and we do this *independently of its level.* (Piaget et al., 1977, vol. II, p. 303, emphasis added.)

When one comes to this statement in Piaget's summary at the end of the second volume on the specific topic of reflective abstraction, it becomes clear that the sequence in which he usually discusses the three types is a little misleading, because it is neither a developmental nor a logical sequence. What he rightly calls "reflective thought" and lists as the second of three types describes a cognitive phenomenon that is much more sophisticated than reflective abstractions of type one or type three and, moreover, is relevant also as a further development of empirical abstraction.

I would suggest that the two meanings of the word "reflection" be assigned in the following way to Piaget's classification of abstractions: it should be interpreted as *projection* and *adjusted organization* on another operational level in the case of reflective abstraction type one and pseudo-empirical abstraction; and it should be taken as *conscious thought* in the case of reflective abstraction type two (also called "reflected").

In his two volumes *La prise de conscience* (1974b) and *Réussir et comprendre* (1974c), there is a wealth of observational material from which Piaget and his collaborators infer that consciousness appears hesitantly in small steps each of which conceptualizes a more or less specific way of operating. Like von Humboldt, Piaget takes the mind's ability to step out of the experiential flow for granted, but he then endeavors to map when and under what conditions the subject's awareness of its own operating sets in; and he tries to establish how action evolves in its relation to the conceptualization which characterizes the attainment of consciousness (Piaget, 1974*b*, p. 275ff). In the subsequent volume, he provides an excellent definition of what it is that awareness contributes.

> To succeed is to comprehend in action a given situation to a degree sufficient to attain the proposed goals: to understand is to master in thought the same situations to the point that one can resolve the problems they pose with regard to the why and the how of the links one has established and used in one's actions. (Piaget, 1974c, p. 237)

The cumulative result of the minute investigations contained in these two volumes enabled Piaget to come up with an extremely sophisticated description of the mutual interaction between the construction of successful schemes and the construction of abstracted understandings, an interaction that eventually leads to accommodations and to finding solutions to problems in the re-presentational mode, i.e., without having to have run into them on the level of sensory-motor experience.

In this context, one further thing must be added. In the earlier sections, I discussed the fact that re-presentation follows upon recognition and that the "pointing" function of symbols follows as the result of familiarity with the symbols' power to bring forth re-presentations that are based on empirical abstractions. As the examples I gave of abstracted motor patterns should make clear, symbols can be used simply to point to such patterns, in which case the re-presentation of action can be curtailed, provided the subject has consciously conceptualized the action and knows how to re-present it.

I now want to emphasize that this pointing function of symbols makes possible a way of mental operating that requires conscious conceptualization and, as a result, gives more power to the symbols. Once reflective thought can be applied to the kind of abstraction Piaget ascribed to Aristotle (cf. passage quoted above), there will be awareness not only of what is being operated on but also of the operations that are being carried out. Piaget suggested this in an earlier context:

> A form is indissociable from its content in perception but can be manipulated independently of its content in the realm of operations, in which even forms devoid of content can be constructed and manipulated. ...logicomathematical operations allow the construction of arrangements which are independent of content ... pure forms ... simply *based on symbols*. (Piaget, 1969, p. 288, emphasis added.)

In *my* terms, this means symbols can be associated with operations, and once the operations have become quite familiar, the symbols can be used to point to them without the need to produce an actual re-presentation of carrying them out. If this is accepted as a working hypothesis, we have a model for a mathematical activity that was very well characterized by Juan Caramuel,[13] twenty-five years before Locke published his *Essay*:

> When I hear or read a phrase such as "The Saracen army was eight times larger than the Venetian one, yet a quarter of its men fell on the battlefield, a quarter were taken prisoner, and half took to flight", I may admire the noble effort of the Venetians and I can also understand the proportions, without determining a single number. If someone asked me how many Turks there were, how many were killed, how many captured, how many fled, I could not answer unless one of the indeterminate numbers had been determined. ...
>
> Thus the need arose to add to common arithmetic, which deals with the *determinate numbers*, another to deal with the *indeterminate numbers*. (Caramuel, 1670/1977, p. 37, original emphasis.)

In Europe, Caramuel says, this "other" arithmetic, which deals with abstractions that are "more abstract than the abstract concept of number," became known as *algebra*. Given the model of abstraction and reflection I have discussed in these pages, it is not difficult to see what this further abstraction resides in. To produce an actual re-presentation of the operative pattern abstracted from the arithmetical operation of, say, division, specific numbers are needed. This is analogous to the need of specific properties when the re-presentation of, say, an apple is to be produced. But there is a difference: The properties required to form an apple re-presentation are sensory properties, whereas the numbers needed to re-present an "operative pattern" in arithmetic are themselves abstractions from mental operations. Yet, once symbols have been associated with the abstracted operative pattern, these symbols, thanks to their power of functioning as pointers, can be *understood*, without the actual production of the associated re-presentation--provided the user knows how to produce it when the numerical material is available.

Conclusion

Abstraction, re-presentation, reflection, and conscious conceptualization interact on various levels of mental operating. In the course of these processes, what was produced by one cycle of operations can be taken as "given content" by the next one, which may then coordinate it to create a new "form," a new structure; and any such structure can be consciously conceptualized and associated with a symbol. The structure that then functions as the symbol's *meaning* for the particular cognizing subject may have gone through several cycles of abstraction and reorganization. This is one reason why the conventional view of language is misleading. In my experience, the notion that word/symbols have fixed meanings that are *shared* by every user of the language breaks down in any conversation that attempts an interaction on the level of concepts, that is, attempts to go beyond a simple exchange of soothing familiar sounds.

In analyses like those I have tried to lay out in this chapter, one chooses the words that one considers the most adequate to establish the similarities, differences, and relationships one has in mind. But the meanings of whatever words one chooses are one's own, and there is no way of presenting them to a reader for inspection. This, of course, is the very same situation I find myself in, vis-a-vis the writings of Piaget. There is no way of discovering what he had in mind--not even by reading him in French. All I--or anyone--could do, is to "interpret," to construct and reconstruct until a satisfactory degree of coherence is achieved among the conceptual structures one has built up on the basis of the read text.

This situation, I keep reiterating, is no different from the situation we are in, vis-a-vis our non-linguistic experience, i.e., the experience of what we like to call "the world." What matters there is that the conceptual structures we abstract turn out to be suitable in the pursuit of our goals; and if they do suit our purposes, that they can be brought into some kind of harmony with one another. This is the same, whether the goals are on the level of sensory-motor experience or of reflective thought. From this perspective, the test of anyone's account that purports to interpret direct experience or the writings of another must be whether or not this account brings forth in the reader a network of conceptualizations and reflective thought that he or she finds coherent and useful.

Philosophical Postscript

It may be time for a professional philosopher to reevaluate the opposition between empiricism and rationalism. The rift has been exaggerated by an often ill-informed tradition in the course of the last hundred years, and the polarization has led to utter mindlessness on the one side and to various kinds of solipsism on the other. Yet, if we return to Locke from the partially Kantian position of a constructivist such as Piaget, we may be able to reformulate the difference.

> 24. *The Origin of all our Knowledge*. -- In time the mind comes to reflect on its own operations about ideas got by sensation, and thereby stores itself with a new set of ideas, which I call ideas of reflection. These are the impressions

> that are made on our senses by outward objects that are extrinsical to the mind; and its own operations, proceeding from powers intrinsical and proper to itself, which, when reflected on by itself, become also objects of comtemplation--are, as I have said, the origin of all knowledge. (Locke, 1690; Book II, Ch. 1,§24)[14]

With one modification, this statement fits well into my interpretation of Piaget's analysis of abstractions. The modification concerns, of course, the "outward objects that are extrinsical to the mind." In Piaget's view, exogenous and endogenous do *not* refer to an inside and an outside of the organism, but are relative to the mental process that is going on at the moment. The "internal" construct that is formed by the coordination of sensory-motor elements on one level becomes "external" material for the coordination of operations on the next higher level and the only thing Piaget assumes as a given starting-point for this otherwise closed but spiraling process is the presence of a few fixed action patterns at the beginning of the infant's cognitive development.

Both Locke's and Piaget's model of the cognizing organism acknowledge the senses and the operations of the mind as the two sources of ideas. Locke believed that the sensory source of ideas, the "impressions" generated by "outward objects," provided the mind with some sort of picture of an outside world. Piaget saw perception as the result of the subject's actions and mental operations aimed at providing not a picture of, but an adaptive fit into the structure of that outer world. The functional primacy of the two sources, consequently, is assigned differently: Piaget posits the active mind that organizes sensation and perception as primary, whereas Locke, especially later in his work, tends to emphasize the passive reception of impressions by the senses. The difference, however, takes on an altogether changed character, once one considers that the concept of knowledge is not the same for both thinkers. For Locke it still involved the notion of Truth as correspondence to an independent outside world; for Piaget, in contrast, it has the biologist's meaning of *functional fit* or *viability* as the indispensable condition of organic survival.

The difference, therefore could be characterized by saying that classical empiricism accepts without question the static notion of *being*, whereas constructivist rationalism accepts without question the dynamic notion of *living*.

Notes

I am greatly indebted to Les Steffe and John Richards for their extensive critique of my manuscript on the same topic, circulated in 1982, and I thank Cliff Konold and Charlotte v. G. for comments on a recent draft of parts of this paper.

1. Locke (1690), Introduction, 1.

2. McClellan & Dewey (1895), p. 27.

3. Locke divided this work into Chapters, Books, and numbered paragraphs.

4. Memory, as Heinz von Foerster (1965) pointed out, cannot be a fixed record (because the capacity of heads, even on the molecular level, is simply not large enough); hence, it must be thought of as *dynamic*, i.e., as a mechanism that *reconstructs* rather than stores.

5. A first English translation of von Humboldt's aphorisms was published by Rotenstreich (1974). The slightly different translations given here are mine.

6. It is crucial to keep in mind that Piaget emphatically stated that knowledge could not be a *copy* or *picture* of an external reality; hence, for him, "imitation" did not mean producing a replica of an object outside the subject's experiential field, but rather the regeneration of an externalized experience.

7. Cf. William James (1892/1962; p. 227): "One of the most extraordinary facts of our life is that, although we are besieged at every moment by impressions from our whole sensory surface, we notice so very small a part of them."

8. Ceccato said this several times in our discussions on the operations that constitute "meaning" (1947-1952, when I was the translator for his journal *Methodos*).

9. Note that my use of the word "symbol" is not the same as Piaget's for whom symbols had to have an *iconic* relation to their referents.

10. Needless to say, the perturbation may simply lead to a retrial or to a modification of the activity; the latter, if successful, would also constitute an *accommodation.*

11. Having introduced an idea from Ceccato's operational analyses into Piaget's model, I today believe that the motion necessary in perception need not be physical, but can often be replaced by the motion of the perceiver's focus of attention (cf. von Glasersfeld, 1981).

12. The title of his volume, as Leslie Smith (1981), one of the few conscientious interpreters of Piaget, pointed out, was mistranslated as "The grasp of consciousness" and should have been rendered as "the onset" or "attainment of consciousness."

13. I owe knowledge of Caramuel's work to my late friend Paolo Terzi, who immediately recognized the value of the 17th century Latin treatise, when it was accidentally found in the Library of Vigevano. Caramuel, a Spanish nobleman, architect, mathematician, and philosopher of science, had been "exiled" as bishop to that small Lombard city because he had had several disagreements with the Vatican.

14. It may be helpful to remember that the first sentence in Kant's *Critique of pure reason* (1781) reads: "Experience is undoubtedly the first product that our intelligence brings forth, by operating on the material of sensory impressions."

5 A Pre-Logical Model of Rationality

Mark H. Bickhard

The nurturance of rationality is among the fundamental goals of education, but one that is not likely to be effectively pursued on the basis of prevalent notions of rationality. In particular, rationality is sometimes wittingly, more often unthinkingly, equated with logic, and the nurturance of rationality is correspondingly equated with the teaching of valid logical rules. I wish to argue that logic cannot be a valid explication of rationality and that teaching logical rules cannot effectively nurture rationality

Rationality as Logic

Perhaps the simplest perspective on the inadequacy of the "rationality equals logic" position is via the recognition that valid logics cannot construct logics more powerful than themselves; in any practical sense, they cannot create new logics at all. If rationality *is* logic, then the historical development of logic itself *and* the development of logic (= rationality) in the individual are both intrinsically *non*rational. At best, they become matters of evolutionary or historical accident, or perhaps mere matters of rhetoric, thus fundamentally undermining their claims to have any basic validity. In such a view, any alternative such product of accident or rhetoric has just as much claim to validity as does our current understanding of logic and rationality, and a rampant relativism is unavoidable. An impasse is reached at which there is little left of rationality that can claim to be rational.

Misconstrued notions of rationality, however, are not important just for ultimate philosophical groundings. They are also crucial for understanding, and guiding the teaching of rational domains of inquiry - which embraces philosophy, logic, mathematics, science, ethics, and esthetics, among others - and for nurturing the development of rationality in individuals. Models of rationality that are implicitly accidental or rhetorically based do not draw on or connect with anything intrinsic in the person or in thought. Correspondingly, they inevitably tend toward relativism (or, more likely, authoritarianism) in their approach to instruction. By definition, if there is no rational grounding for rationality, then it must be given a *non*rational grounding or else be given no grounding at all.

I wish to sketch an alternative model of the nature of rationality that does provide for its own rational grounding and development. Furthermore,

I will indicate how the central rational domains of logic and mathematics fit within this alternative view. This view of rationality provides for its own rational grounding by explicating rationality as a development and specialization of properties that are already necessary and intrinsic in thought. Rationality is no longer a set of standards or norms imposed on thought from outside, thus having no grounds *within* thought, but emerges inherently in the properties of thought itself. Most fundamentally, it is not just that the development of thought happens at some point to evolve "rationality," but, more deeply, that there is an intrinsic *concern* in thought which, when developed, *is* rationality. The grounding for rationality, then, is in that concern, and rationality is the development and specialization of that concern.

Such a grounding will provide both a motivational and an epistemic grounding for rationality. It will be neither a purely epistemic model that cannot address the question "Why should anyone care?" nor a purely motivational model that cannot address the question "What epistemic grounding can it claim?" Instead, both questions must be answered simultaneously. I will argue, in fact, that both questions, within the right framework, are ultimately the same.

Some Thoughts About Thought

Clearly, to make good on such a promissory note, the first requirement is for a defensible model of thought, or at least a model of sufficient properties of thought to support the relevant arguments. The purportedly sufficient model of thought is elsewhere called interactivism (e.g., Bickhard, 1980a; Bickhard and Richie, 1983; Campbell and Bickhard, 1986), but it will not be developed here. Instead, I will approach the issues via the sufficient properties of thought, providing independent plausibility arguments for each of these properties before proceeding to the model of rationality which they yield. These properties are, in fact, intrinsically necessary to the nature of thought and representation according to the interactivist model, and that model, in turn, is arguably a necessary model of thought and representation per se. Space and the technical nature and complexity of some of the relevant arguments, however, preclude a full presentation here (for a brief introduction, see Bickhard, chap. 2).

There are two basic properties of mentality that form the foundation for the interactive model of rationality to be presented here: 1) learning and the development of thought are intrinsically constructive processes of trial and error, of variation and selection; and 2) thought involves the inherent potentiality of reflection, of conscious thinking about thinking.

The arguments for the variation and selection nature of learning and development are basically those for the *existence* of variation and selection processes *and* those for the *elimination* of purported alternative models. For existence: variations and selections clearly are established as evolutionary processes. They are straightforwardly observable in any history of a dispute, say, in science, where new proposals are first put forward and later eliminated - selected out - by logical or empirical criticism; and they are present in any developmental progression in which errors occur and are eliminated. To claim that variation and selection processes exist, in fact, is not particularly controversial.

The arguments for the elimination of alternatives to variation and selection, however, are less well known, particularly for the alternative known as induction. The first alternative to be considered is really just a denial in disguise of learning or development, and is thus easily eliminated, at least insofar as it is correctly recognized: the supposed "alternative" is prescience or foreknowledge. This position contends that there is neither learning nor development, but that all possible knowledge is already available. However, because learning and development certainly seem to occur, this position faces serious prima facie implausibility. Nevertheless, in the contemporary version of innatism, such a position has recently gained much ground. A *valid* innatist model is simply a model of how some specific knowledge arose, or could have arisen, by evolutionary processes. Such models have important roles to play in understanding how humans and other animals function and develop. Innatisms, however, that proceed by arguing that some or all of knowledge *cannot* develop, and therefore that it must be all already present (innate) are fundamentally arguing that such knowledge cannot come into being at all (Bickhard, chap. 2; Campbell and Bickhard, 1987; Fodor, 1975, 1981; Piattelli-Palmarini, 1980). If such arguments are at all valid, then such knowledge cannot come into being via evolutionary variation and selection any more than it can by developmental variation and selection, so innatism is not an available solution. Such models require an external agency to insert something akin to knowledge of Plato's forms at some point in evolution or in development. Such models are not scientific, they are supernatural.

The primary alternative to variation and selection as the process of creating new knowledge is the supposed process of induction. Induction is the impressing of patterns from the environment into the mind, thereby giving the mind knowledge of those patterns. Induction is also a presumed form of "rational" justification for such knowledge. The notion has fared badly in both of its forms as origin *and* as justification of knowledge, but I am primarily concerned here with induction as origin. The argument against it is quite simple: in order for a mind to notice such a pattern in the environment, the possibility of that pattern must be already epistemically available to that mind; otherwise, it would not in fact be noticed. If mind must bring the possibility of that pattern to the world as a kind of hypothesis to be tested, then we have a version of variation and selection after all, not simply a passive induction of knowledge (Popper, 1959, 1965, 1972, 1985).

By both existence and elimination arguments, then, the development of thought and representation must occur via some forms of variation and selection processes.

Arguments for the inherent potentiality of reflection are in some senses simpler than for variation and selection, and in certain other senses more complex. They are *simpler* in that such a possibility is prima facie not in question; any reflection on the issue at all already settles the issue. At best it might be argued that certain aspects of thought are not available to direct reflection, but that too is not seriously in question (e.g., neurochemical foundations) and does not alter the basic argument to be presented. The issues are more *complex* in that, however incontrovertible the *fact* of reflection is, it is extremely difficult to *account* for such consciousness or, for that matter, to even characterize what it is. Most models, together with programmatic frameworks for modeling such as contemporary information processing approaches, not only ignore consciousness in fact, but are fundamentally incapable of modeling it in principle (Campbell and Bickhard, 1986). Perforce, they cannot provide guidance for understanding the nature of, or for the nurture of, anything that intrinsically involves consciousness such as, in particular, rationality. Interactivism does claim to model the nature and the evolution of consciousness (Bickhard, 1980b; Campbell and Bickhard, 1986), but again, only the potentiality for conscious reflection will be needed in the ensuing argument, and that is established in the very asking of the question.

The Nature of Rationality

The model of rationality that emerges from these two properties is simply that of *reflective* epistemic variation and selection, or meta variation and selection. It is this notion that will be presented and elaborated.

Variation and selection involves both processes and principles of variation and processes and principles of selection. But because all evolution and development ultimately rest on variation and selection, any particular constructive variation processes must themselves be the products of previous variation and selection constructions. The same holds true for the selection process and principles, but there is a deep asymmetry between the relationships between successive constructive variations and the relationships between successive selection processes and principles.

New *constructive* variation processes are those that have *satisfied* whatever new *selection* principles might have been constructed. These may be, in some sense, variants on previous such constructive processes, or they may have no carry over or accumulation of content at all. The positive contents of variational constructed processes can, in principle, change unpredictably and radically from one version to the next. To deny that possibility is, in effect, to revert to prescience: to claim that some parts or aspects of current knowledge are, in fact, ultimately true and need never be changed. Common sense and the previous argument against prescience, as well as the long history of science, belie such notions.

Successive principles of *selection* on the other hand, of *criticism* when rendered in language, always have a very special relationship with preceding principles of selection. In particular, new critical principles always *apply* to earlier such principles. They may apply in such a way as to affirm earlier principles (perhaps strengthening their grounds or broadening their scope); or they may apply in such a way as to *in*firm earlier principles (perhaps undermining what was taken to be a support or presenting a counterexample to a claim); or they may apply so as to affirm and infirm simultaneously (for instance, if a previous principle were to be differentiated into two or more special versions with differing scopes of application, thus infirming the earlier claimed general scope, but affirming the basic common form of the principle). Fundamentally, however, higher-level critical principles are *about* lower-level, earlier, critical principles. It is this *aboutness* that intrinsically requires the potentiality of reflection. Conversely, given a variation and selection process together with the possibility of reflective consciousness, critical principles arise simply as the lifting of variation and selection (especially selection) into the realm of conscious process.

The Necessity of Rationality

Rationality in this sense of the progressive construction of critical principles, and of the succession of positive contents and constructive processes that successively satisfy them, is a necessary tendency of the development of thought. That there is an inherent developmental tendency to progressively construct critical principles follows immediately from the explication just given. Insofar as variation and selection are intrinsic to mind, and insofar as the possibility of conscious reflection is intrinsic to mind, then *reflective* levels of variation and selection are also intrinsic potentialities. But I am proposing that reflective variation and selection, with the intrinsic primacy of the principles of selection, i.e., critical principles, *is* rationality. A tendency for the development of hierarchies of critical principles, then, is intrinsic to thought, via its intrinsic characteristics of variation and selection constructivism and potentiality for reflection.

Furthermore, internal principles and processes of selection are intrinsically motivational. At the lowest level, they are the processes involved in appreciation of success, failure, pain, hunger, loss, and so on. At higher levels of reflection, critical selection principles are not just relative to the environment, but they are *about* the properties and processes and even the very constitution of the overall system or its parts. They are satisfied, or fail to be satisfied, by the person and by his or her various aspects as a person. Satisfaction of these principles is *sought* via instrumental and self-constitutive efforts. Such selection principles, then, do not just function mechanistically, they constitute the basic carings and concerns of the person. As proposed in Campbell and Bickhard (1986), they constitute the basic values of a person, which, in turn, constitute the core of a person's identity, and form the leading edge of further development. It is nontraditional to consider *epistemic* values, as is being proposed here, instead of ethical values in the narrow sense (which are fundamentally "just" epistemic values about ethical issues, or ethical principles that "fit" those epistemic values), but that is simply an unfortunate legacy of the Kantian heritage in which ethics is separated from the fundamental concerns of the person (MacIntyre, 1981). Critical principles, then, as values, as higher order goals, and as higher order selection principles, function deeply and intrinsically in the motivational carings and concerns of persons.

The Rationality of Rationality

However much the rationality of critical principles may be an intrinsic tendency and motivational involvement of persons, it remains to inquire as to the self-consistency of this notion of rationality. In what sense is the development of a hierarchy of critical selection principles a rational development? Is that sense consistent with the very model of rationality being proposed? The answers to these questions are again simple in principle once some misconceptions are eliminated. Most commonly, such questions are asked with the presupposition that any answer must involve a model of the justification of the "rational" products, and that such justification, in turn, involves some demonstration of absolute or relative movement toward Truth. The paradigm case is again that of induction, which is supposed to provide greater and greater assurance or probability of truth the larger the range of data or experience upon which it is based, and is assured to converge on Truth in the limit. Models of movement toward Truth have universally suffered failure. The reasons for such failure have most often been taken to be either technical, and therefore correctable by a technical fix, or at least correctable by some modified version of the notion of justification. In some "radical" attempts, some other positive content is substituted for Truth as the ultimate epistemic goal-pragmatic success or problem-solving success, for example (e.g., Laudan, 1977, on scientific rationality).

The issues involved here are too ramified and complex to address thoroughly, so I will suggest my own diagnosis of this class of failures. The contended solution then follows readily. The basic diagnosis is that standard approaches to the nature of rationality are defined in terms of some sort of positive content such as truth, pragmatic success, and so on. Hence the internal consistency of such models must be gauged in terms of movement toward a totality or purity of such contents or in terms of accumulation of such contents. Unfortunately, we cannot know what those limit points are or what part of our current contents partake of the "correct" contents: to already know that would constitute prescience. But without knowing in advance what the correct content should be, or at least in what "direction" it should lie, it becomes impossible to ensure that rational development is in fact progressive according to the purported definition of rationality. Without such assurance, however, in at least a probabilistic form, such models of rationality cannot support their own claims to rationality: they are reflexively inconsistent.

The solution is already inherent in these explications. The rational progressivity of rationality lies not in accumulating positive content, nor in moving toward positive content, but instead in moving *away* from *error*. The basic character of variation and selection is not to ensure a particular content in the surviving constructions, but to ensure in some minimal sense the avoidance of error, of eliminative selection. Variation and selection processes progress in their capacity to satisfy selection criteria. From the standpoint of meta variation and selection, the hierarchy of critical selection principles *constitutes* the individual's (or society's) knowledge of what sorts of errors can be made and should be avoided. Progressive knowledge is knowledge that satisfies more elaborated and advanced critical principles, that avoids more sophisticated and advanced sorts of error. This model of rationality guarantees a progressive tendency by the criterion of movement away from error. Rationality as meta variation and selection, then, *does* address its own self-consistency, its own selection criteria for rationality, and it satisfies them.

The Rationality of Necessity

As one implication of the critical principle model of rationality, I will consider *necessity*. Necessity, in its various forms, partakes of the intrinsic tendencies toward rationality. There are intrinsic tendencies toward necessities and, more deeply, toward the development of critical principles for necessities.

One of the fundamental functions of selection criteria is to encounter *exceptions* to the current system, cases or instances in which the system fails. Construction of further variations then attempts to avoid such exceptions, and further critical principles attempt to anticipate such exceptions. That further constructions attempt to avoid such exceptions derives directly from the nature of variation and selection. That construction of further critical principles attempts to anticipate such exceptions follows from realizing that critical principles stand in the stead of potential eliminative selections from whatever the system interacts with or is about. *Satisfying* critical principles is in the service of *avoiding* relevant lower level selections. Encountering an exception, then, is not only a failure of the system, it is also a failure - at least of scope, if not of content - of the selection principles which that system is taken to satisfy.

The development of critical principles, then, involves an intrinsic tendency toward processes and representations without exceptions - toward *exceptionlessness*. But exceptionlessness *is* necessity. Having no exceptions across some domain of potential exceptions constitutes what we mean by necessity, and differing such domains provide differing varieties of necessity: pragmatic necessity, physical necessity, logical necessity, moral necessity, existential necessity, and so on. Furthermore, as such criteria differentiate and specialize as distinct critical principles, not only are necessities (as properties of system and representation) tendencies of rationality, but values - critical principles - of necessity likewise are part of that intrinsic developmental tendency.

Necessity is a limiting case of one aspect of rationality - the aspect of having no exceptions. It is a limiting case not only in the sense of constituting a point beyond which no further improvement is possible - no exceptions is the best there is with respect to this criterion - but also in the sense that there can never be any absolute assurance that that point has been reached. For a system to not be selected against by a principle like necessity requires only that no exceptions or likely exceptions be currently known; it does not guarantee that no exceptions will be discovered in the future. Not being currently selected against by a principle is not the same as ultimately satisfying that principle. There is an asymmetry here that is deeply similar to Popper's distinction between empirical confirmation of a scientific theory and falsification of a theory: no amount of accepted confirmation is compelling, while even a single accepted falsification is compelling. It is the same asymmetry at a conceptual and metaconceptual level (empirical falsification can be viewed as eliminative selection against an empirical criterion, an empirical value).

This asymmetry is important because it allows for the rational employment of critical principles as selection criteria that we can never be assured are, in fact, satisfied - that we can never, with philosophical certainty, *believe* to be satisfied. Many important epistemic values - such as necessity, truth, realism, and so on - partake of this asymmetry, and are correspondingly impossible to incorporate into positive belief-focused explications of rationality. There can never be enough reason to believe, say, that contemporary physics describes reality (especially given the ultimate failures of even the best supported physics of the past), but it will nevertheless still be rational to apply realism as a critical principle against current theories: discovering their failures to meet the demands of realism *is* an increase in knowledge, and leads to further constructive improvements.

The Nurturance of Rationality

The dialectic between variations and selections proceeds apace, inherent in the nature of life and of thought. The origins and developments of explicit *meta*variations and selections, however, thus of explicit rationality, are only an intrinsic potentiality and tendency, one that may or may not be actualized or that may be actualized to varying degrees. The critical principle model of rationality yields some distinct implications concerning the conditions that might encourage and nurture that development.

The first implication is a straightforward extension of the model to the social realm. Social norms of openness to criticism tend to elicit attempts to discover criticisms and also to anticipate and avoid potential criticisms; they tend to lead, therefore, to the discovery of critical principles and of ways of thought that seem to satisfy or to fit those principles. This point would seem to hold just as much historically, as with the ancient Greeks and their prodigious contributions to philosophy (Annas, 1986), as it holds for family or classroom interactions.

A second implication concerns the relationships between rational domains and the applicable critical principles. Knowledge of the contents of such a domain *without* knowledge of the relevant critical principles within which those contents are presumed to fit is knowledge without motivation or understanding. The critical principles provide answers to the "Why bother?" and "Why this way?" questions that make rational knowledge rational. Instruction that ignores such concerns is intrinsically, even if inadvertently, authoritarian and rote. Too often, of course, children (*and* adults) are left to discover the relevant principles on their own. Commonly, such material simply remains arbitrary, irrelevant, and boring.

A third implication is, in effect, a deeper corollary of the second. The hierarchy of critical principles for a rational domain is intrinsically historical; it is an implicit sedimentation of the rational historical development of the field. Correspondingly, a powerful perspective on any rational domain will be the historical perspective on its development. To understand what problems were being addressed, what assumptions were being made, and what critical constraints were being observed in creating the field (including those that overthrew previous conceptualizations of problems, assumptions, and constraints) is to know the field as a vital domain that the individual can connect with and potentially participate in. The historical perspective is grievously absent in contemporary conceptions of education, and, even when it is present, the critical principles involved are rarely explicated. More typically, the historical succession of positive contents is simply recounted, leaving the student with another arbitrary, irrelevant, and boring collection of items.

The fourth and last implication that I will address follows from the point that although rationality is an intrinsic *tendency* of thought and develops from an intrinsic *concern* of thought, rationality is *not* an intrinsic *part* or *aspect* of thought per se. Rationality is an epistemic and inferential self-disciplining of thought, a *meta*-development of the selectional aspect of thought, not the essence of thought. This view contrasts, for example, with the view in which rationality is equated to logic and logic is taken to be (inherent in) the rules of thought. In this view, and in others in which rationality is a "part" or domain or form of thought, rationality is conceived of as being in opposition to the "passions." The implication that follows from the interactive model, in contrast, is that rationality emerges out of creative engagement with problems, perhaps even ludic or fascinated (passionate) engagement. In other words, the standard opposition between creativity, emotions, curiosity, and so on (the passions), on the one hand, and the discipline or domain of rationality, on the other hand, is a false opposition. The two sides of the supposed opposition are in fact the aspects of the constructivist variations and the motivated selections, respectively, that are intrinsic to *all* of thought. *All* such aspects will be evoked when learning proceeds within the framework of motivated problems; problems, in turn, will *be* motivated insofar as they connect with the current knowledge and values (epistemic and otherwise) of the person involved. Knowledge emerges because someone has reason to care or to be interested; this model suggests that that is the most powerful manner in which rationality will develop as well.

Conclusion

Contemporary notions of rationality tend to be explicated in terms of the rational contents of positive knowledge. This leads to both philosophical impasses and educational distortions. Philosophically, for example, such notions cannot account for the rationality of their own development. In terms of positive contents, such as rules of logic, rationality would have to have already been in existence in order for it to have come into existence in a rational manner. Educationally, the nurturance of rationality becomes bare instruction in those contents, which leaves them unmotivated and not understood. These points are deeply intuitive once made, but insofar as they have conflicted with dominant notions of rationality, those intuitions have had no guidance. Rationality as critical principles dissolves such impasses and resolves such educational distortions.

6 Recursion and the Mathematical Experience

Thomas E. Kieren and Susan E.B. Pirie

Mathematics for children, as they 'do' it and build onto their own knowledge, is a complex phenomenon. As with personal knowledge of mathematics for any person, it is multifaceted, a view supported by mathematicians (e.g. Davis and Hersh, 1986) and psychologists interested in mathematics education (e.g. Vergnaud, 1983). There are many ways to observe and hence say something about children's mathematics. One could consider mathematical thought mechanisms used by children. One could analyze specific aspects of the complexity. However, in this essay we try to observe and talk about children's mathematical experience as a complex whole.

The authors have observed this complexity in their own work. Kieren (1988) has suggested an ideal structure of personal mathematical knowledge. Such knowledge is seen as an integrated structure including ones own mathematically related knowledge built up through everyday life experience (ethno-mathematical knowledge). A second kind of mathematical knowledge (intuitively built knowledge) arises from the conjoint use of thinking tools such as counting, or equi-division; imagery or figural items; and the informal use of language, which may however be standard mathematical language. In an ideal structure, the two informal knowledge types above would form a basis for technical symbolic mathematical knowledge, which is derived from more formal thoughts or actions and from patterns of symbolic transformations. Finally, there is one's mathematical knowledge built from deductions within a system of axiomatic assumptions. One should of course not think of a single mathematical knowledge system for each person, but many inter-related systems, each of which could have components from the four layers above. It is pertinent to ask here how the components from these layers might be inter-related for any individual and how such a knowledge structure might be viewed as a growing whole.

It is evident that if one's mathematical knowledge structure is to form an entity rather than a fragmented collection of information, one would have to possess understanding. Pirie (1988) has suggested that describing different kinds of understanding (instrumental, relational, procedural, concrete, etc.) as a means of differentiating people with differing kinds of understanding is inadequate for capturing the meaning of understanding exhibited by children. Understanding mathematics is itself a complex phenomenon involving, for the person doing it, many different aspects. What is needed is an insightful way of viewing the whole of a person's growing mathematical understanding built using their own knowledge.

It is our intention in this essay to use recursion as a metaphor to look at the phenomenon of personal mathematical knowledge and understanding. Recursion has been selected on two grounds. First, it is a way of describing complex phenomena in which the whole at any time is structurally similar to, but not reducible to its previous states. Second, with Maturana and Varela (1980) we see human beings, including children, as knowers of mathematics, as prime examples of autopoietic systems. That is, children are self-referencing and exist in their sphere of behavioral possibilities with the primary conservative goal of self-maintenance. One element of their sphere of actions is their own mental states and actions. Since children are self-referencing, a significant means of cognition is recursive in nature; that is, knowing occurs through thought actions which entail the results of previous thought actions as 'inputs'. Thus, in looking at the complex phenomenon of the whole of a person's mathematical knowledge and mathematical understanding, recursion appears to be an appropriate metaphor.

What is Recursion?

In discussing recursion we follow Vitale (1989) who notes that:

> ...there are no problems, no objects specifically recursive; it is the way we tackle them and represent them that makes them recursive.

Thus in what follows we make no claim that either the mathematical knowledge of children, or the mathematical activities of knowledge-building and problems solving are inherently recursive. We are simply observing them from a recursive point of view. What we do argue is that such an approach is possible and furthermore it is instructive in understanding children's mathematics, in specifying mathematics for children, in giving further insight into knowledge and understanding in mathematics (Pirie, 1987) and in giving insight into the possible roles of child-child and teacher-child communications within the children's mathematical experiences.

While recursion does have a specific technical meaning in many computer procedural languages, it is another matter to precisely define recursion in a manner useful in the study of mathematical activity of children. Vitale (1989) has attempted to define recursion in a way which he claims to be useful in psycho-cognitive research.
This definition is as follows:

a relation A = (. . . A) is recursive if:

a. (self-reference) its right hand side contains in some way the same entity 'A' that is present in the left hand side;

b. (level-stepping) There is an element of change. 'A' is present in the right hand side in a somewhat different aspect than in the left hand side.

This definition is useful in what follows in several ways. It is designed to look at a whole relation as recursive, as opposed to looking at the details of particular steps. It also highlights two features which will prove important in our metaphorical use of recursion; non-reducible levels and self-similarity in structure. Because one has a history of experience in mathematics, one's knowledge and understanding at any point in time should entail previous knowledge and understanding. Thus it will be useful to look at such knowledge in some temporally 'levelled' way. Further, as will be argued later, since mathematical knowledge building and understanding should be a growing process, any state of such knowledge should have structural similarity to previous states.

When applying this definition of recursion to the examining of children's mathematical experience, it is useful to explore further the levels of such experience. One could, without nominally violating the constraints of the above definitions, view a new level of mathematical knowledge as simply an embroidery of the old. This would suggest that one's current mathematical knowledge could be 'reduced' to one's previous mathematical knowledge. That is, one would be tempted to use a task analysis approach to 'higher' knowledge reducing it to its 'lower' component levels. We, however, choose to take a stronger position on change of knowledge level through recursion. Margenau (1987) sees the change in scientific knowledge as 'transcendence with compatibility'. In terms of recursion we take this to mean that a new level of knowledge transcends but is compatible with the knowledge previously held. One level of knowing cannot be simplistically reduced to a previous level; it is neither reducible to the connected sequence of its parts nor reducible to any one of them. From this perspective the mathematical knowledge of a child at a particular point in time is a transcending whole which non-destructively integrates previous states or levels of knowing and in particular can call these previous states in problem solving. While the actual actions of this integration are not discussed in detail here, it is assumed that various kinds of transformations and accommodations are involved [see Steffe (chap. 3) and Bickhard (chap. 2)].

In using recursion as a metaphor for looking at a child's mathematical experience, we will highlight the fact that the mathematical knowing, understanding or problem solving activity of a child:

- is self-referencing;
- involves states which differ from one another;
- is such that the structure of the states in a particular knowledge building or problem solving activity are self similar;
- that a current state of mathematical knowing transcendently elaborates previous states, and integrates or entails these states in the sense that they can be called into current knowing actions.

We will use these highlighted features of recursion in our analysis and will call the recursion thus defined *transcendent recursion.*

Because this construct, transcendent recursion, is to be used to look at children's mathematical activity, several other constructs need elaboration before we apply the metaphor. The first is 'abstraction'[1]. We claim that human beings and in particular the children and young adults between the ages of seven and nineteen described below are capable of abstract thought. For Maturana and Varela (1980) this means that individual can distinguish between ideas seen as products of their own mental actions and those which appear to them to be driven by outside sources. Evidence for such abstraction is found in reactions to the following task.

> Given two equal collections of chips (A & B), if one or more (n) are taken from A and added to B, which set then has more, A or B, and how many more?

Cooper (chap. 7) notes that children from age seven on can cope with this task (and even learn the n -- 2n generalization) while younger children have difficulty. It would seem that the younger child is acting simply on her observation and reacting to the environment, while the 7 year old is acting on her own idea (abstraction) of the environment. This view is congruent with Piaget(1980) who sees a change from exogenously governed knowledge to endogenously constructed knowledge in children in the early school age range.

In forming such abstractions and building their own mathematics, children use language. Following Maturana and Varela (1980), we claim that language has an orienting function. In other words, language does not carry information from speaker to listener, but can serve an orienting function for the latter. In particular, language allows a child to orient herself to her own actions, particularly mental ones, thus facilitating abstraction and the recursive use of her own knowledge in the building of new patterns of knowledge. To do this, to use her own ideas in order to fashion a new idea, it is necessary that the existing knowledge may be personally internalized, be part of her understanding and not merely a reaction to external stimuli. In a sense a child with this kind of knowledge is ready for further, more elaborate knowing or doing and language use can facilitate this.

Secondly, 'level-connectedness' is both a condition and consequence of recursion. Although a child may use a new higher level piece of knowledge in doing mathematical tasks without reference to other knowledge, the high level knowledge can call knowledge at a lower level if necessary. This suggests that a child can 'drop' to a previously developed level of mathematical experience to construct or construe the basis for an elaboration on a higher level experience. Perhaps more importantly, a child can validate 'new' knowledge - that is, show herself that it is true - by calling on previously held knowledge. Examples of several different kinds of level-connectedness are given in the analysis below.

Finally, as suggested by both Vitale (1989) and Margenau (1987), high level knowledge must be compatible with lower level knowledge for understanding to take place. It appears that a necessary, but clearly not sufficient, condition for a piece of knowledge to transcend its antecedents is that it carry forward some central elements of the form, process or patterns of the antecedent ideas. It is this which allows such transcendent knowledge to be, itself, easily elaborated. That is, if an individual's knowledge is a transcendently recursive whole, then she is potentially conscious of the building of that knowledge.

What Contributes to Children's Mathematical Experience?

Having provided a rough definition of recursion which might prove useful in considering the mathematical experience of children, we now turn to this experience. We propose to look at two complex facets of mathematical experience: problem solving and knowledge building. Problem solving and knowledge building allow us to specifically sample mathematical experience occurring in non-routine, non-topic-related circumstances as well as such experience in building knowledge within a more content-oriented school syllabus. In addition, we will cross-set these two experience domains against two uses of the recursive lens: recursion within the actual activities of children which are constituting parts of their mathematical experience; and mathematical experience in general as a recursive phenomena. In the first case we will look at the transcripts of young persons doing mathematics. We will try to metaphorically use the elements of transcendent recursion as a lens to see and understand growth in these activities. The second use of our recursive lens is in an attempt to understand a piece of mathematical knowledge or a problem solving act as a *dynamic whole*. We are not using recursion as a scientific description of the change between levels of knowledge or as a replacement for abstraction and its underlying transformations. We are using it to understand problem solving, knowing, and understanding as wholes entailing such abstractions and their resulting substance and form.

Recursion in student's thoughts and actions within the problem solving experience

The following well-known problem was posed to a class of thirty-five adolescent students:

> How many handshakes are needed so that each person in this room shakes hands with each other person exactly once?

Students worked in groups and many of the activities reported in the problem solving literature were observed: specializing, making tables of results and special cases, and representing the problem symbolically or geometrically and attacking it in this form. The whole process of problem solving as recursion will be discussed later, but here two protocols are analyzed using the elements of transcendent recursion described above.

Group 1 (reporting results to the whole class)

Joanne: Think of the persons as being somehow in order, like in terms of distance from the door. The person furthest from the door walks along shaking hands and counting the handshakes. When you reach the person closest to the door you shake hands and report your handshake count to the last person and leave. Now the person second furthest from the door does the same, and there you have it.

Simon: (Elaborating, perhaps transcendently on Joanne's idea) There's a way to know how far they were from the door - one for the closest; thirty-five for the furthest. Now number thirty-five shakes thirty-four hands and reports the count to one and leaves the room. Numbers thirty-fours shakes thirty-three hands, reports and leaves, and so on. Person three shakes two hands, reports and leaves. Person two shakes person one's hand and leaves. Person one takes out his calculator and adds $34 + 33 + 32 + \ldots + 1$, and goes out and reports.

Teacher: (Hoping to provoke some generalizations) What about a room with 1000 people? With n people?
Students: It's just the same.
Teacher: What is "it's"?
Alison: Well you just do exactly the same thing.

The process as a whole contains at least two elements of transcendent recursion. The process is self-referencing in its actual action and in the students' own assessment of it (see Joanne). There are obvious levels within this problem and the whole solution does not reduce to any or all of the levels.

There is a hint of transcendence in Simon's idea which puts the idea of Joanne in a more formal way. Conceptually, the transcendence is seen when Alison realizes that the problem had been solved in general even though no numerical answer had been given. In fact the students never did bother to compute a solution for thirty-five people! One might say that Alison is realizing that generalizing this solution meant noticing that the solution pattern (perhaps as formalized by Simon) would be self-similar at any level. The fact that the problem was never actually 'answered' at any level suggests that the students sense that the solution to the problem does not reduce to an answer or result of a special case, but 'calls' or uses the structure of that special case. The 'structure' of this special case has in it both a substantially correct mathematical idea (a sequence of non-repeating handshakes) and a form by which it can be procedurally described[2].

A second class of fourteen young adults was given the same problem to work on individually. What follows is Jim's response and a brief discussion of it.

Jim: I'd just think of myself as the fourteenth person walking into the room. I'd say "how many handshakes so far?" Then I'd shake hands with the thirteen people and add thirteen to the number already done. That would give the number for fourteen.
Diana: But that doesn't seem right. You don't have an answer.
Shauna: Oh! I see. Yes he does. The thirteenth would have done the same thing - asked "how many?", shaken hands with twelve and added it on. Because by then the twelfth person would have done the same thing and so on down to two persons when there would be one handshake.
Jim: Right. See, the third person would ask and get an answer "one". He'd add his two handshakes to get three and so on back up to me at fourteen.

In the first example above, the students acted in a recursive manner within the problem setting. They "linearized" the situation and suggested use of interactive acts of counting and adding. On the other hand Jim appeared to view the whole problem recursively and produced an informal recursive solution to it.

In using recursion as a metaphorical lens, only some elements are evident in this brief transcript. The notion of a solution in some way referencing itself at different "levels" is evident in Jim's and Shauna's comments. The transcendent nature of the method of solution is best seen in the contrast between Jim's and Shauna's responses and that of Diana. While Diana was seeking an 'answer' for fourteen people, Jim and Shauna appeared to see that what had been done carried the general pattern of the solution of which the actual given problem was only an instance. While there is no direct evidence of this in the above transcript, one again gets the sense that these two see a self-similarity in levels of problem solution. Jim made a particular attempt at the solution; Shauna appeared to see that this local solution embedded the pattern for a general solution. Jim's last remark then specialized to check Shauna's insight.

Problem solving as a recursive activity.

Instead of looking at actual student activity in problem solving through the lens of recursions or trying to precisely identify recursive behavior in problem

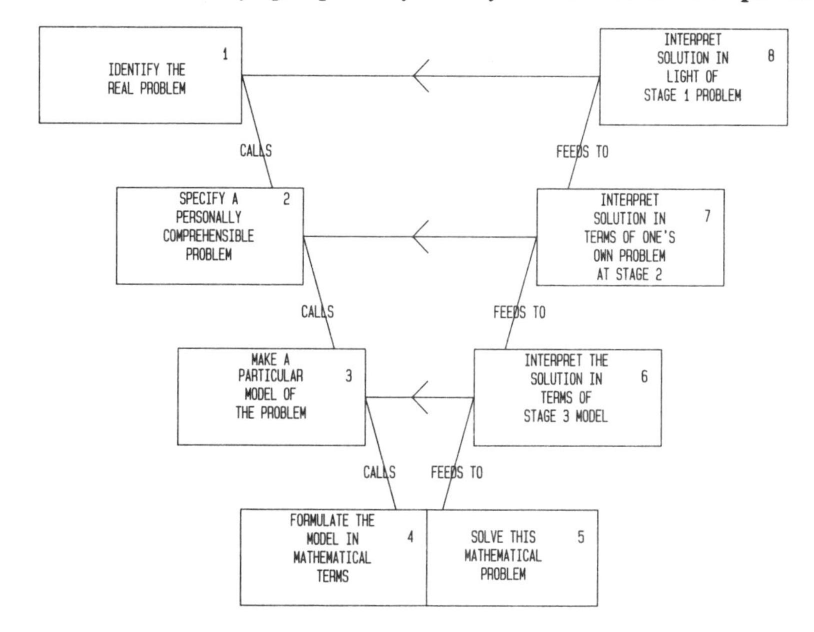

Figure 1. A recursive model of problem solving.

solving one can also look at the whole enterprise of solving a problem from a recursive point of view. What is given below (Figure 1) is an image of this enterprise proposed in Pirie (1987) and modified to highlight some of the recursive features.

Recursion, as we have defined it, is evident in at least four different ways in this view of problem solving. The image is drawn to emphasize the 'drop-down' 'feed-back-up' *levelled* nature of the enterprise. Stages two, three, and four are steps down from the original problem. Each stage represents a specialization of the previous stage. The aim here is personal - *you* want to have a problem (Stage two) which *you* think you should have an image for (Stage three) and *you* can solve mathematically (Stage four). It should be noted here that these stages are paired with corresponding validating, testing stages (1->8; 2->7; 3->6; 4->5). Stages one through four (and particularly two through four) represent different kinds of ways of asking "What in my experience, mathematical, or otherwise, allows me to act?" The validation stages (8-6) correspondingly represent different kinds of ways of asking "what in my experience, mathematical or otherwise enables me to criticize these actions?". Thus, the problem solving act at any stage is self similar involving an action evaluation scheme.

In a global sense this image highlights recursion. The "problems" at stages 2-4 are all deliberately 'copies' of the original, but are different in their specific content, while Stages 5 back to 8 represent the validating of solutions in different ways. There is transcendence in the sense that, for example, the stage 2->7 problem-solution is qualitatively different from and likely more general than the stage 4->5 technical problem-solution. In fact, if you simply solve the mathematical problem at a particular level at stage 5 and are only interested in the answer per se, then problem solving stops. Stages 6-8 require not the input of *an answer*, but the *process or patterns* by which it is attained. Thus this eight-stage image intentionally has the form of a recursive procedure and the general problem solving scheme would function recursively, specializing to reach an effective action, generalizing through evaluating the 'lower' level pattern of action against the criteria of the more general problem.

Figure 1 could also be looked at as a series of inter-related triangles of action and evaluation as seen in Figure 2 below. If your problem solutions fails to be valid at any of the stages 6-8, then you re-examine the step from the corresponding action stage. This action is illustrated by the arrows pairing Stages 6 with 3, 7 with 2, and 8 with 1. For example, failure to validate at Stage 6 requires a re-examination of Stage 3; changes in the model at Stage 3 in turn imply changes at stages 4, 5, and 6. One might say that this triangular sub-set of actions within this model represents, the 'recursive fabric' of the problem solving enterprise. Since problem solving has the feature of levels, together with conservation of the essence of the problem, a difficulty at any one level induces changes at other levels. Further, the recursive fabric indicates that the problem solving

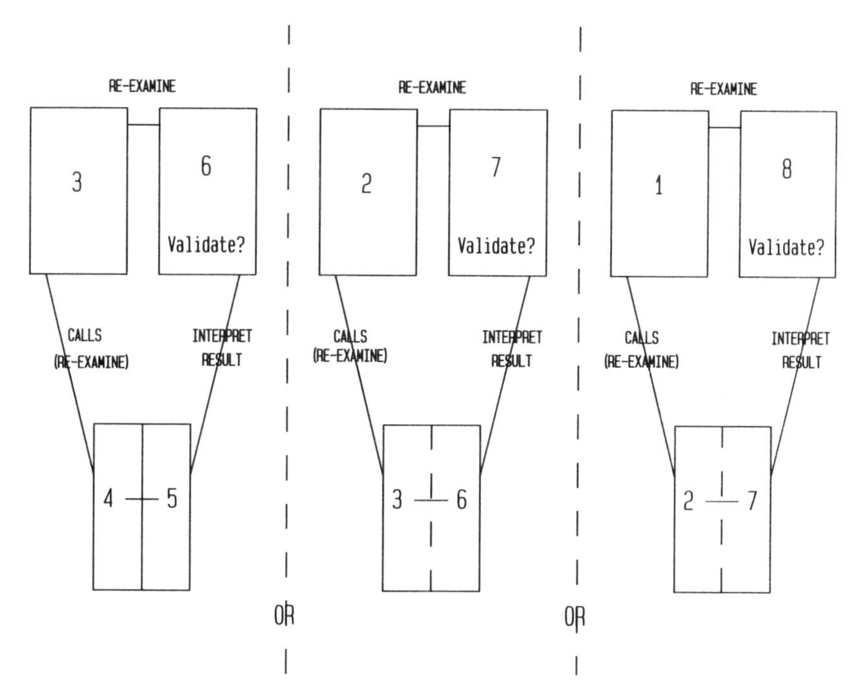

Figure 2. Self similar action-reflection triangles in problem solving.

process for a person might contain several passes through the same stage, each one altering the problem at that stage and its contribution to later stages in the process.

Looking at Stages 4 and 5 specifically one sees what might be called 'locally recursive activity'. This involves the mathematical specializing and generalizing as described in detail by Mason (1982) or by Pirie (1987). This activity is local in the sense that its goal is the specialized one of setting and solving a specific mathematical problem.

Even then particular procedures to derive particular results (e.g. computation) might involve this recursive structure. Cockcroft (1988) describes the action of an adult worker who when asked to compute 7 x 98 responded as follows: *"Work out 3 x 98, add the result to itself and add 98".* (The worker did not know how to multiply by seven). One can easily view this act recursively (cf. Figure 3). Thus even in looking at computations through this lens one can see calls to other stages of prior knowledge and the related evaluations.

The whole structure and any of the specializing-validating 'triangles' at any stage are self-similar. Each stage or level corresponds to a different

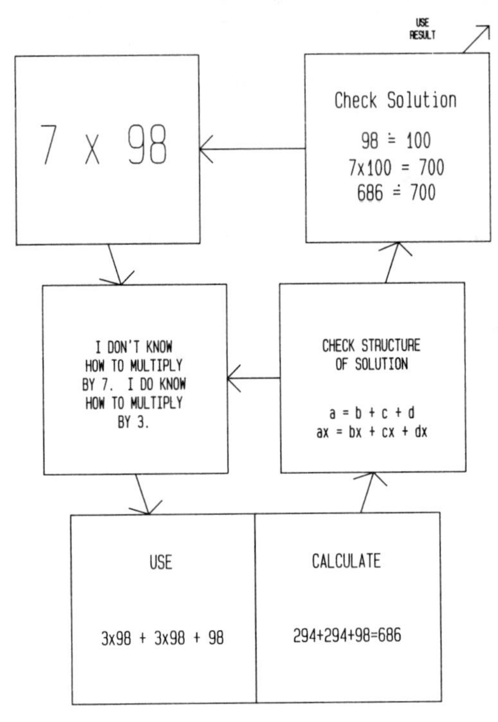

Figure 3. A recursive analysis of a computation.

qualitative aspect of the problem (personal question, personal model or cause, mathematical model, specific mathematical example, computation, . . .) showing the level stepping nature of problem solving. In the sense that the higher levels (1 or 2 compared to 3 or 4) represent more general stages, problem solving has a transcendent recursive structure.

This argument can be carried further. If you have a solution to the problem validated at Stage 8 you could stop or you could generalize further the original Problem 1 to Problem 1'. The transcendent recursive nature of such a choice is illustrated in the model below.

This image in the form of a helix (Figure 4) is meant to highlight that Problem 1' is a transcendent elaboration of Problem 1, not only in terms of generalizing Problem 1. It also calls not for using previous solutions but, for repeating the solution *process* (Stages 2' through 8') based on the solution process already obtained for Problem 1 (Stages 2 through 8). This general transcendent view of problem solving is like the helical view of mathematical thinking in Mason (1982, pp. 156, 157). It also contains a further illustration of the recursive fabric of problem solving in that the formulations of the personally comprehensible problem at 2' may well be a transcendent

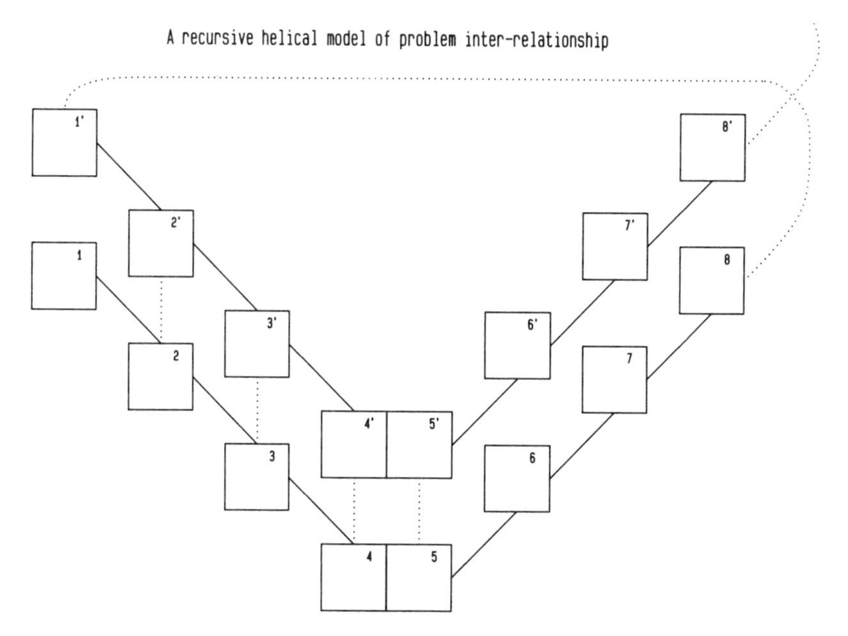

Figure 4. A recursive helical model of problem inter-relationship.

elaboration of the personal problem at Stage 2. The solution process for the new problem does not depend on dropping to the related stages of the prior related problem for the actual solution. Rather, the connections between the two cycles of the helix suggest that work at related stage of a previous problem is a source for a *pattern* of action or evaluation in the current problem.

We offer here one example illustrating the problem solving process from these recursive viewpoints.

Stage 1: Why do we see rainbows?
Stage 2: Why is light split into an array of colors by water? (personal question)
Stage 3: Suppose the drop of water is a rectangular prism, in fact a cube. (personal model)
Stage 4: Use laws of reflection, and refraction to get appropriate equations for this problem. (mathematical model)
Stage 5: Solve and check the solution mathematically.

Stage 6: The form and substance of the Stage 5 solution fits with the Stage 3 model. (solution - model match)

Stage 7: The solution is meaningless at Stage 2, perhaps because water drops are not cubical. (model - personal question mis-match). Look at the call from Stage 2 to Stage 3 and change the model at Stage 3. Imagine, for instance, that a water drop is a sphere. Go around Stages 3 to 7 again. The solution now appears reasonable for the Stage 2 question. (model - personal question match)

Stage 8: Interpretation of solution gives an answer to Stage 1. (solution - original problem match)

Now, one could stay at this level of the helix and generalize by asking if there might be a more precise problem (Stage 2a) (why is the array of colors always in the same order?) or a more precise model (Stage 3a), (suppose rain is composed of tear shaped drops) either of which in turn would affect all the subsequent stages. Alternatively one could transcend the previous problem and enter a new level of the helix (in this case by interchanging the refractive media).

Stage 1 'Can a fish see 'airbows'?'

The above example of problem solving illustrates its levelled dropping down, climbing-up nature. It also is meant to illustrate the recursive fabric and the level change evaluation level-recreation activity in the stage 2 - 3 - 6 - 7 'triangle' (cf. Figure 2). Finally, it illustrates the beginning stage of the helical transcendent elaboration of one problem by a second.

Recursion in students' thoughts and actions within the knowledge building activity.

The action of children actually engaged in developing standard mathematical ideas can also be studied using the notions of recursion developed above. What follows are excerpts from transcripts of children working on tasks relating to conceptualizing and operating on fractions.

In the first instance, three seven year-old girls, Hanne, Anne, and Nan are working on this mathematical task: Two groups of children are sharing some pizzas. In Group 1, seven children are sharing three pizzas and in Group 2, three children are sharing one pizza. Each child in each group gets a *fair share*.

How much does child A get?
How much does child B get?

Who gets more or do they have the same?
Why do you think that?

Group 1

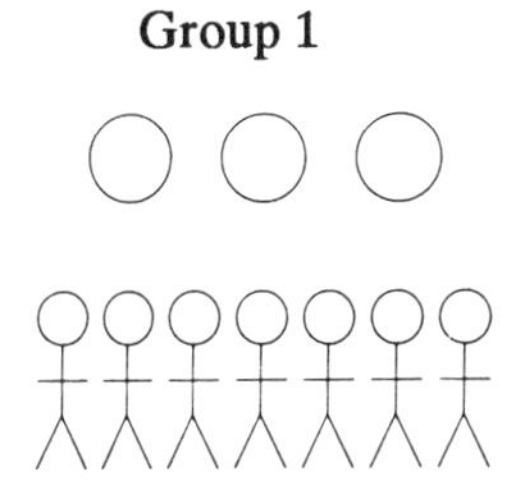

Group 2

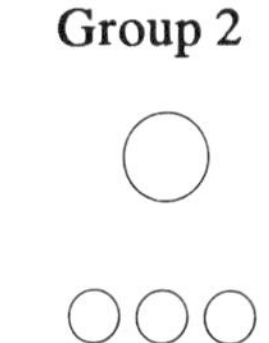

This task and protocol parts are based a study of children's language by Wales (1984) in which she worked with 7 to 9 year-olds in a city in Western Canada.

Hanne: A is hard - let's skip it. Nan (pause while looking, sketching etc. goes on).

Hanne: B is easy, you 'Y' it

(a discussion ensues in which the teacher finally persuades Hanne to show Anne and Nan how to actually do what she is doing. In doing so Anne and Nan start simply applying the letter Y to the circles representing the pizza.)

Hanne: Let's use 'Ys' on A.

(Action 1, draws: 

)

In Action 1, Hanne cuts the three pizzas into "fair shares" and gives one third to each.

(Action 2, 

4 5 6

)

Then she cuts the remaining two thirds into seven smaller pieces.

Hanne: Oh, I see! A gets a third and a bite. A gets more.
(the other girls are totally puzzled by these actions).

Hanne herself is initially confused by the problem of sharing three pizzas among seven and does not immediately see breaking into sevenths, or indeed any other solution. Although she uses idiosyncratic language ("You 'Y' it") she observes that B is easy to handle. One might say her language orients her to and is also put for the act of dividing into three equal parts. Her comments to the other girls (as well as her actions on other tasks demanding partitioning into three parts) leads one to believe that Hanne has abstracted "one third" at least in this local positive sense. This idea is hers and not simply her mental reaction to the world. One might also think of her "Y" as a formalizing of "thirding". She was replacing a meaningful thought action with its form or sign.

She then appears to rethink the 'three for seven' problem in terms she already knows. Although she cannot divide three wholes into seven parts directly, she calls her form for thirding into play. This enables her to see the remaining two pieces as together breakable into sevenths.

The critical thing to note here is that Hanne made an abstraction which she formalized. She was ready to call this form as part of a new scheme for dividing up. Her scheme for partitioning three into seven equal parts was clearly 'the same' as her scheme for one into three parts, in that it involved physical quotients. But it was different in that it used an expanded notion of quotient and quantity (the sum of two quantitative actions). Further, it was clearly *connected* to her previous abstraction. Evidence in later problems showed that it continued a pattern of conceptual actions which she used in other circumstances. On the other hand, although Anne and Nan were 'taught' how to partition into three, they did not have this action as a useable abstraction. It might be said that they had been taught a form before they had a meaningful action themselves. Thus the 'Y' form could not replace for them a quotient action since it did not exist. In a sense they could act, but not understand or use their action because it was prematurely formal. Since their idea of one-third as a quotient was not yet a mental item for them, it could not be used recursively in new ideas or settings.

One can also contrast the action of Hanne, which we are observing as transcendent recursion, with that of three eight year olds; Robert, Dan, and Kath as they work on a similar task below.

Robert: Oh A is easy
A gets three pieces (See action on 'A' below)
Dan & Kath: (nod in agreement and appear to think the sharing is right)
Dan: B is hard, though.

All three get to work on their own copies of the problem and proceed in the manner illustrated below:

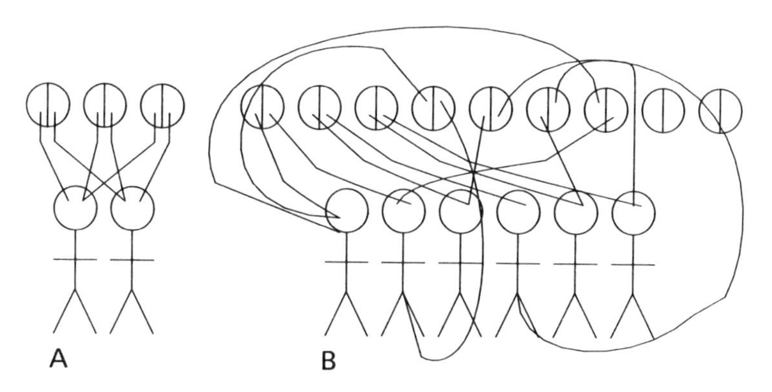

Soon all are hopelessly lost and all but Robert give up.

Now the actual halving action seems known by all and is likely applied because there are two persons to share the three pizzas. Although Robert, Dan, and Kath now repeatedly use the form of the halving action, they do not appear to see it as part of a quantitative scheme. There is no recursion even though their actions on 9 for 6 use halving. It is merely the 'blind' repetition of halving, not the use of the same form in a new way. There is no level stepping or transcendence. Here simply knowing half and its form is no help in problem solving or engendering a new scheme of action.

The examples above are observations of single learning episodes from the point of view of recursion. However, recursion can also prove useful in discussing the development and structure of a whole piece of personal mathematical knowledge. What follows is the description of the development of the notion of division of fractions by one eleven year old girl, Katie. The discussion is based on the analysis of an extended transcript of Katie working with a friend, Anna over a period of a week, as well as that of an interview which took place some weeks after the initial learning. This transcript forms part of an extensive project on classroom discussion and has been described in detail elsewhere (Pirie, 1988).

Katie's knowledge-building sequence below occurred in the context of a classroom in England where pupil-pupil class discussion was a highly used phenomenon. Prior to what is described below, the class had come to the conclusion that, at least in a local way, $1 \div 1/n = n$.

The teacher then posed the question: What is $1 \div 2/3$? Discussing and using this image,

the class concluded that: $1 \div 2/3$ is 1.

One could summarize Katie's (and her classmates') knowledge of division of fractions at this level by the question:

Level 1 "How many 2/3 sized pieces in a whole?"

Rather than try to disentangle or discuss the use of the remaining piece compared with "2/3 as a unit", the teacher wisely suggested that the class work on the problems 2÷2/3, 3÷2/3, 4÷2/3, 5÷2/3. Some of her classmates simply used a replication of the previous visual image:

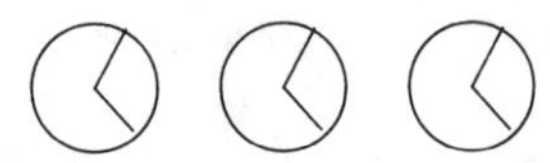

and concluded that the "answers" were 2, 3, 4, and 5. At the point of 3÷2/3 illustrated above, Katie suddenly thought of 2/3 as an abstracted, constructible, divisible quantity. She informed Anna that they could use the "other bits" to make up one and a half more two thirds and concluded that

3÷2/3 = 4 1/2.

A summary of her knowledge at this point is:

Level 2 "How many abstracted a/b sized pieces fit into another quantity?"

Notice here that her Level 2 knowledge was like, and in fact called, her Level 1 knowledge (self-reference). But it was clearly transcendent and division of fractions was answering another but structurally similar question (level-stepping)

Katie was now able, in this locally abstract but image-guided way, to answer a variety of fraction division questions. At one point in this activity she was working on the task 6 1/3÷2/3 and commented that the actual drawing of the circles and the 2/3 pieces was "boring". She suggested that "6 wholes would have 6 times 3 . . . 18 thirds and half that number of two-thirds. So, 6 1/3 had nineteen thirds and 9 1/2 two-thirds." Noting a pattern in her thinking Katie tried 2 1/4÷2/3. "Times by 3" produced 6 3/4 ("third pieces") and then "half of it" gave 3 3/8.

Here 2/3 was being treated both as quantity and as a semi-symbolic entity. Katie did break it down into the 3 (thirds) and the 2. She still thought of her results as a quantity and it is as if her use of symbols was an analog to an imagined action. At this level Katie's division of fraction knowledge can be summarized by the question:

Level 3 "What multiple, described semi-symbolically, of the quantity a/b fits into to quantity c/d?"

One should note that for Katie '2/3' has itself again changed character. It is still an abstracted quantity, but it has become more mathematical in nature. Katie appeared to use its components symbolically as analogs for particular actions. Again, Level 3 knowledge used Level 2 knowledge; but in a transcendent way. Level 3 knowledge is now on its way to fractional number knowledge; Level 2 knowledge was about quantity in a very 'physical' way. Using Level 3 knowledge Katie seemed to consider the form of Level 2 actions and substitute formal numerical activities in a piece-wise way for them.

Some weeks later Katie was interviewed and asked what division of fractions as all about. She replied, "Oh, you just turn it upside down and multiply". When asked if she could do the processes she easily did the following:

$$2/3 \div 1/5 = 2/3 \times 5/1 = 10/3 = 3\ 1/3$$

One might now summarize Katie's knowledge in the statement:

Level 4 Division of fractions is a process on two mathematical entities or numbers.

At this point we can only surmise that this knowledge (multiply by the inverse) is an abstraction and contraction of the form of the earlier knowledge 'multiply by 3 and take half'. It appears that 2/3 and 1/5 are examples of a new kind of entity now very much like numbers rather than a quantity.

After doing the above calculation Katie was asked if she could show how it worked. At this Katie drew the following:

This is quite simply a picture of the symbolic sentence. At this point one might conclude that her Level 4 knowledge was simply computational and not connected to the knowledge prior to it.

The interviewer changed the form of her questioning slightly. Interviewer: "Will $50 \div 1/3$ be bigger or smaller than 50?" (an order of magnitude question). This time Katie did NOT "turn it upside down and multiply". Katie: "Bigger, because a third (quantity), will fit into 50 (quantity) quite a lot of times". Here one can notice Katie calling Level 2 knowledge.

Interviewer: Will 4/5÷1/2 be bigger or smaller than 1?

Katie: Bigger than! because the 4/5 is more than 1/2. If it was going to be 1 or smaller it would have to be smaller than two-and-one-half fifths. You've got to see how many times a half goes into 4/5. If it was two-and-one-half fifths it would be 1 and if it is less it is still going to go in less, but if it is more that two-and-one-half, it is going in more than one times.

Notice that while Level 4 knowledge could have been used, Katie now seemed to want to show herself how it worked. That is, she needed to validate this knowledge *as true for herself.* To do so she appeared to use Level 3 reasoning - such reasoning is language based, but clearly is analogous to actual quantitative actions.

To summarize, one can view the whole of Katie's knowledge of division of fractions recursively. Knowledge at each level appears to be a transcendently changed copy of the previous level. The knowledge appears to be a connected whole for Katie. That is, she could function with ease at any level, but when necessary she could call a prior level. As well as the process of division at one level transcending the prior one, the ideas and form of fractional quantity or number also appears to be an abstraction of the prior level. Finally, for Katie, the knowledge at one level seemed to use in some way the process or operational contraction of it. It is almost as if the substance of knowledge at one level in some way used Katie's formalization of knowledge at a previous level.

Mathematical Knowledge Building as a Recursive Activity

Maturana and Varela (1987) see cognition as an effective action by a human being that will enable her to continue to live in a definite environment and "bring forth the world". In the previous section we have observed Hanne and Katie bringing forth a mathematical world for themselves and we have tried to show how the concepts related to recursion enable such observation. In what way can one use these concepts in observing the whole enterprise of constructing a piece of mathematical knowledge? The holistic levels of knowledge building growth can be diagrammatically represented as in Figure 5. Recursion is seen to occur as thinkers move between levels of sophistication. Each level is contained within succeeding levels, each boundary is dependent on the forms and processes within and constrained by those without. More specifically, the division of a fraction knowledge of an individual can be represented by a magnification of the action at the boundary between intuitive and symbolic thinking (Figure 6).

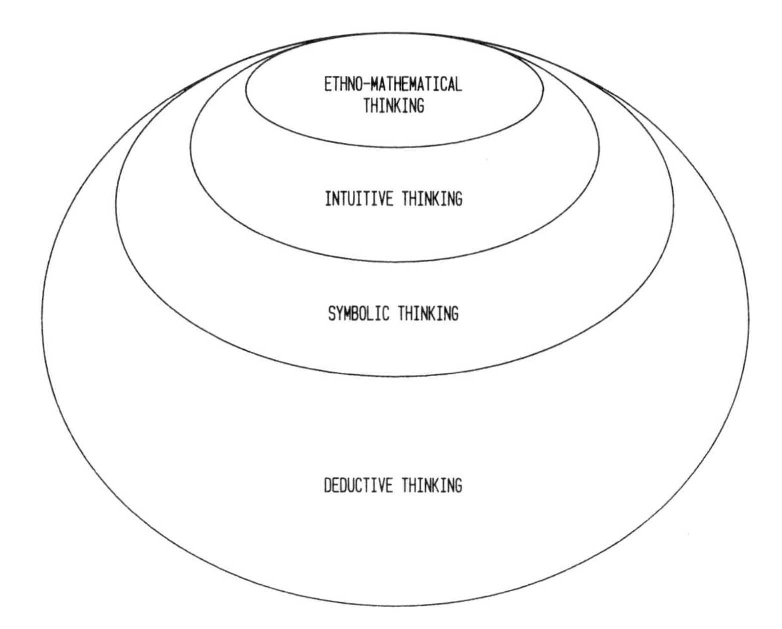

Figure 5. Mathematical knowing as a recursive model

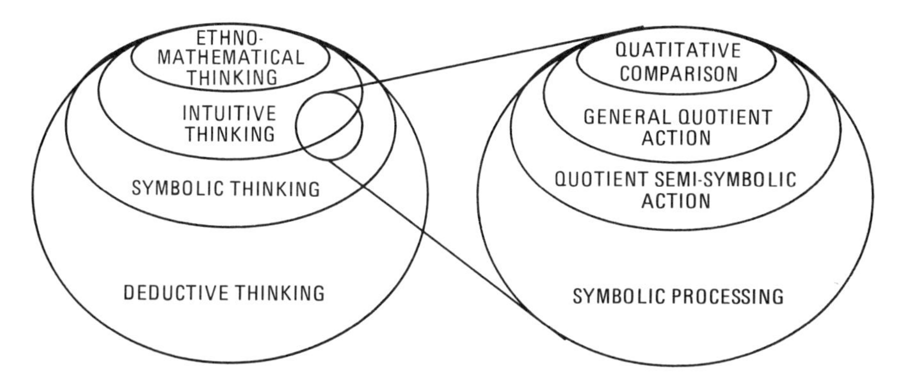

Figure 6. The knowledge building of a person moving from intuitive to symbolic thinking.

We observe that this knowledge is growing self-similar knowledge; it has recursive integrity. *"Symbolic processing"* both integrates, and is in a way very much like but transcends (both in its act and its field of activities) *"quotient semi-symbolic action"* and so on. One might say that the levels are self-similar in that they are all about the question "how many a/b pieces are in the quantity c/d?". Yet there is clearly level-stepping and transcendence. Levels QC and QA represent intuitive knowledge while level QS is knowledge in transition to technical symbolic knowledge and level SP is the achievement of the latter, again showing a levelled self similarity with the model of knowing as a recursive whole in Figure 5. Yet as illustrated in Katie's responses to the fractional division 'order of magnitude' questions, Level SP knowledge can call prior levels or patterns of knowing into play.

The path Katie took in building up this whole knowledge system is illustrated in Figure 7 below.

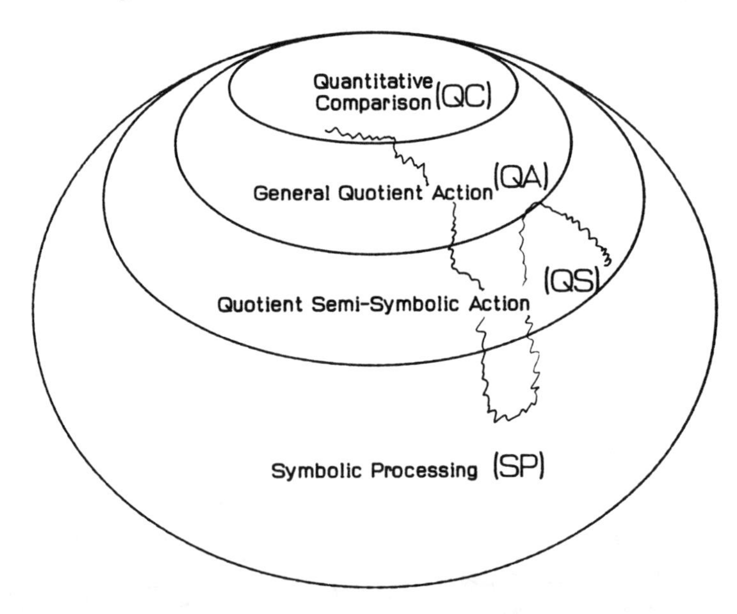

Figure 7. The non-linear recursive knowledge building path of one individual.

As seen in the transcript, this path is not linear. Prior knowledge levels are revisited in building up new knowledge levels. Even when symbolic process knowledge is available and used, when faced with a new situation, Katie calls or revisits more physical action oriented knowing (QA) in reconstructing semi-symbolic (QS) knowledge about the order of quotients of fractional numbers.

We might more generally hypothesize that in a recursive personal knowledge system knowledge building does not simply occur from the "inside-out", but can occur in any order among closely related knowledge levels.

Finally it should be noted that this image represents a growing whole and should not be thought of as a sequence. As can be seen in the work of Katie or Hanne, a new level of knowing appears to call and grow out of the old in an almost continuous manner. Further, though for them the new level of knowing is more powerful (or easier to use) than the prior ones, the prior knowledge exists as useful in its own right and can be called in the face of new questions. Thus it is misleading to think of QS knowing as 'lower' than SP knowing.

We would like to compare this recursive knowledge structure as illustrated by Katie with a sequential structure of knowledge for the addition of fractions as traditionally taught in the classroom (Figure 8 below).

Traditional Level 1

A fraction arises from counting parts in a single pre-divided whole.

Traditional Level 2

Fractions in general are double counts: count the total number of parts (denominator) then count the number of shaded parts (numerator).

Traditional Level 3

Fractions with like denominators are added by counting like parts.

Traditional Level 4

Fractions in general are added by number theoretically finding common denominators.

Traditional Level 5

Standard symbolic addition of fractions.

Figure 8. A traditional non-recursive levelled sequence of adding fractions.

This traditional sequence does not have recursive integrity. To be sure each higher level transcends or goes beyond its predecessor. But at key points this knowledge is not connected. Knowledge from a lower level is only nominally and not actively integrated by the next. Thus Traditional Level 2 is based on counting. It is a double count on a static phenomenon and in a sense emphasizes the ratio notion of fractions (as well as the part of a single whole notion). Traditional Level 3 involves counting but not ratios. Single parts or pieces are counted. The real discontinuity comes at Traditional Level 4 where adding is based on number theoretic notions and only nominally in the end involves counting like parts or pieces. In the traditional sequence we have non-recursive phenomenon. In particular, Traditional Level 4 is not really the same as Traditional Level 3. Thus it is not surprising that persons asked to learn fractions in this sequence ordinarily flounder. Such knowledge is not a coherent whole, but a sequence of isolated pieces. For those who do not see the number theoretic 'trick' there is no way back to previous knowing. Even for those who do the symbolic manipulations correctly, there is no way of validating their knowing or of dropping back to prior levels when new fractional knowing is needed.

Summary

It was our intent to look at the phenomena of mathematical experience using recursion as a metaphor. In so doing we used a concept of recursion which maintained the form of the mathematical or computer science notion but went beyond it. We maintained the self-referencing and level-stepping features, since these features allow recursion to be used as a metaphor for phenomena that might have the features of self-similarity and levelled complexity. In expanding the term to 'transcendent recursion' we emphasize that the recursive relation (and hence the phenomena studied through it) as a whole does not reduce to a particular level nor is it in any way a simple replication of a level.

In looking at mathematical experience we considered the complex phenomena of personal mathematical problem solving, and knowledge building with understanding, both in terms of actual activities of children and general models. It was assumed that personal knowledge and problem solving were growing action-oriented personal phenomena. Hence transcendent recursion might prove to be a useful metaphor in looking at them.

One feature of both problem solving and knowledge building revealed under a recursive lens is that although they are complex and locally idiosyncratic phenomena, they also exhibit a dynamic regularity. At any stage of the problem solving activity, or when considering the activity as a whole, a self-similar structure is evident. We envisioned this structure diagrammatically as a triangle of activities involving specializing and creating results and generalizing through interpretation and validation. This suggests that in teaching and evaluating problem solving this interrelated complex of actions is a critical feature. Further, any such triangle could, if necessary, call through specialization such a triangle at another level of specificity. The whole problem solving activity could also be regarded as an open ended helix, each level of which represents a transcendent problem and its solution approach. This provides a different way of looking at the well known heuristic advice "have you solved a similar problem". As suggested in the discussion of Figure 3, activity at any problem solving stage can be modelled after that of the related stage of a prior problem in the helix.

With respect to knowledge building and understanding (which the recursive lens suggests are inter-related), the dynamic regularity is evidenced in what we term recursive integrity. That is, at any level of knowing, knowledge is essentially about the 'same' thing. Of course later stages might be more symbolic in character, but because of this level similarity, a person with such symbolic knowledge, can validate it and the results it produces against what is for them a real manifestation of that knowledge at a prior level.

In recursion, each level, while embedding previous levels and using results from them, has its own status. Similarly, in a personal growing historical knowledge structure, each level, which embeds and calls prior levels, has its own status, and thus, if we talk of levels of knowledge, whilst one level may be temporally prior to another, or indeed prior in general mathematical development, it should not be thought of as a 'lower' level of knowledge. Similarly, mathematical understanding, is not reached at any particular level. In a structure with recursive integrity a person is always exhibiting a degree of understanding.

We would therefore argue that the recursive lens does orient us to new insights into the complex phenomena of mathematical experience whether in non-routine problem solving or personal mathematical knowing and understanding.

Notes

1. Our use of abstraction is different from, but compatible with, the idea of reflective abstraction used by Steffe (chap. 3) or Bickhard (chap. 2) elsewhere in this volume.

2. It is of interest to note that the substance and form of this solution do not entail any mathematics such as functions, difference tables, or geometry outside that already given in the problem.

7 The Role Mathematical Transformations and Practice in Mathematical Development

Robert G. Cooper, Jr.

The focus of this chapter is on the role of repetitive experience in the development of mathematical concepts. It will be argued that young children can and frequently do discover mathematical properties and invent mathematical rules through experience and practice, substantially independently from direct instruction. Further, it will be argued that a crucial role of repetitive experience is that it can cause the construction and reorganization of knowledge which then provides new information to the developing cognitive system. There are two sources of information that are important for the developments discussed here that are not discovered in this way: subitizing and conventional counting systems. Subitizing or direct perception of small numerosities appears to be innate and does not provide evidence for what is conventionally called the "number concept," nor is it irrelevant to its acquisition, as has sometimes been suggested. Counting is socially transmitted, although I would agree that it is repetitive experience with the counting system that makes it meaningful.

Transfer Problem Illustration

A preview of the types of mathematical development that will be discussed in this chapter can be provided by an illustration from a task adapted from Piaget (1974/1980) that I have used in some of my own research. Performance on this task highlights the role of repeated experience because even adults frequently err until they have at least one experience with the task. These are problems whose solutions are not usually acquired until well after entering school, but they can be considered as similar to the other work discussed in this chapter because the solutions are not usually formally taught. In this task, the child is presented with two equal-numerosity arrays in one-to-one correspondence. One is then screened, and one or more objects are transferred from the visible to the screened array. The child's task is then to predict how many objects must be added to the visible array to make it equal to the screened array. Piaget (Piaget, 1974/1980; Piaget, Grize, Szeminska, & Vinh-Bang, 1977) considered this task to be theoretically interesting because the transfer of n objects produces a difference of 2n because both a subtraction and an addition of n are being performed by the act of transferring the objects. Piaget reported that preschool-age children expect the difference to be n. Somewhat older children learn with repeated examples that a transfer of one produces a difference of two, but they do not generalize this to large arrays that exceed their counting ability. By age 8, experience with a few examples is sufficient for children to correctly predict for all transfers of n objects, and by 9 or 10 they can explain the principle. Our own results have been generally consistent with Piaget's description (Campbell, Cooper, & Blevins, 1983; Cooper, Campbell, & Blevins, 1983).

There are two dimensions of expanding understanding in the transfer problem. One is the increase in the numerosities of the arrays to which inferences can be made. Children initially restrict their inferences to numerosities they can count; they then extend them to large numerosities, and ultimately to abstract quantities. This increase in the range of application of knowledge may seem to be relatively easy to explain, but for young children it is acquired slowly as a consequence of repeated experience. The fact that preschoolers seem to restrict the generalization to countable array sizes highlights the underlying problem: how can one learn that situation A is the same as situation B and hence that the same inference rule applies?

The second dimension of expanding understanding in the transfer problem involves a change in the rule itself, not a change in understanding the situations in which it is appropriate to apply it. Ultimately children can correctly respond that 2n must be added to the visible array for a transfer of n objects, but preschoolers do not acquire this knowledge, despite repeated experience.

Interestingly, among the incorrect rules they generate are an n + 1 rule learned from transfers of 1 and an n + 2 rule learned from transfers of 2. Some children maintain one or the other of the rules in the face of the errors they inevitably produce. Others develop them as restricted rules that apply only to transfers of one numerosity, and hence they would require an infinite set of such rules to handle all possible transfers. In this case, repeated experience with resulting errors has allowed them to develop a correct but limited solution to problems they have encountered with limited generalizability to the potential problem space. One crucial piece of information these children may be lacking is the knowledge that the series 2, 4, 6, ... can be formed by adding each integer to itself, which would be conventionally described as multiplying the set of positive integers by 2.

The features of this illustration which will be repeated throughout this chapter are, first, that the children are acquiring the solution to this problem that makes use of a numerical relationship that they discover themselves. Second, the discovery of this relationship depends on understanding some of the properties of the set of positive even integers or the procedures by which one can construct such a set. Third, the knowledge depends on experience within the task itself and/or prior experience. Fourth, the experience that is relevant involves transformations such as additions and subtractions which lead to movements within a mathematical space.

Plan of Chapter

In this chapter I will present four examples of changes in mathematical concepts that are associated with "extensive practice." The goals of the paper are to use these four examples to illustrate the organization and reorganization of mathematical knowledge. The specific skills to be discussed are infants' acquisition of relative numerosity, preschoolers' understanding of addition and subtraction, preschoolers' understanding of one-to-one correspondence, and elementary school children's acquisition of algebraic concepts. The experiences that are proposed as relevant to development in each of these four areas all involve addition and subtraction. As in practicing typing, repetitive experience is used to mean repeating a set of similar and interrelated activities, not doing exactly the same thing over and over again.

Experience has generally been considered to have two kinds of effects on knowledge: it can create or enhance units of knowledge, and it can reorganize knowledge. The effects of repetitive experience on knowledge, as in studies of the effects of training studies, has almost exclusively been studied from the perspective of creating or enhancing units. When the amount of repetition is one of the variables of interest, the studies are almost always conducted from the perspective of classical theories of learning that suggest that experience in some sense "strengthens" knowledge in the individual. In contrast, in the study of skills, such as changes in typing performance with increased experience, it is found that repetition not only enhances "strength," but it also effects major reorganizations of the skill itself. The possibility that repetition induces similar reorganizations of knowledge, not just of skills, has almost never been seriously considered.

As conceived in this paper, the role of repetitive experience, repeated interaction with the environment, is to generate a richly interconnected cognitive space and/or to reorganize such spaces. In part, the construction of such a space constitutes acquired knowledge and, in part, it can serve as the foundation for reflective abstraction from which new knowledge results. It is for this reason that I use the term repetitive experience rather than practice, since the term practice connotes that the activity being practiced is the skill of interest.

Other effects of repetitive experience

In simple associationist models of the past, repetition was most commonly thought of in terms of increasing the strength of associations. In the stimulus sampling models of Guthrie and Estes, repetition was important because it influenced the probability that any particular aspect of the stimulus array was associated with the appropriate response. The modern PDP (parallel distributed processing) approaches can also be thought of as part of this general approach since the consequence of repeated experience in these models is to adjust the weights in a neural net.

Other roles for repetitive experience in cognitive development have been suggested by other researchers (e.g., Case, 1978, 1985; Shatz, 1978). Case (1978, 1985) has argued for the role of practice in transforming controlled processes into automated ones which require less "M-space." Shatz (1978) suggested that specific practice decreases the cognitive capacity required for specific "information-handling techniques." In both cases, the freed cognitive capacity is thought to contribute to more advanced cognitive accomplishments. Others have made use of Rozin's (1976) concept of accessing and have argued that experience leads to some skills being more accessible and therefore more easily used. Although I do not want to dispute that these changes may occur and that repetitive experience may play a crucial role in such changes, I do dispute the sufficiency of these concepts to account for cognitive development, a point with which at least Case would agree. Thus, although practice may consolidate information or procedures and make them more accessible, this role of practice will not be discussed further in this paper. Instead, the goal of this paper is in part to see how repetitive experience provides the context within which development can occur.

Trabasso--transitive inference

To illustrate how reticent cognitive developmentalists are to consider simple repetition a source of reorganization, consider the work of Trabasso and his colleagues on transitive inference (Bryant & Trabasso, 1971; Riley & Trabasso, 1974; Trabasso, 1975). Trabasso reasonably claimed that children could not be expected to make inferences if they could not remember the premise conditions. Therefore, he gave extensive training on the four premise pairs of a five-item series. In cases where only verbal information was presented concerning the pair relationships, very high attrition occurred among the 4-year-olds, and those who did learn the premise information required an extraordinary number of trials (in many cases more than 200 presentations of each of the four premises). For the select group who were able to learn the premise conditions, many but not all appeared to have organized the information into an ordered series just as an adult would.

The consequences of the very extended training have generally been interpreted as overcoming an "encoding bias." It has been argued that young children tend to treat the information of big and little within a pair as information that one item is big and the other little in some absolute sense (Riley & Trabasso, 1974). A few critics have argued that the nature of the training sequence provided the serial organization which was used when other coding strategies proved inefficient (Blevins, 1981; Breslow, 1981). What I would like to suggest in this paper is that the repetitive experience with the premise pairs required a reorganization within the system for storing the premise information. Unlike Trabasso, I do not want to argue that such a system was there all along, but rather that some children in the study were able to develop such a system, given repeated information about the premise pairs (a position that Gelman and Baillargeon, 1983, explicitly reject). I will not pursue the transitive inference problem further in this paper, but it is a familiar example of the type of problem I wish to address.

Why repetitive experience has been ignored

Before moving to the four examples of the development of mathematical concepts, two preliminary issues will be addressed. First, reasons why repetitive experience has not been considered in the emergence and reorganization of knowledge will be examined. Then an example of the formation of cognitive maps will be used to provide a quick illustration of how the role of experience will be treated in this paper.

Reorganizational consequences of experience have generally not been addressed for two reasons, one historical and one theoretical. Historically, such questions concerning repetition have for decades been considered in the purview of classical learning approaches that have no theoretical way of addressing organizational issues at all. Theoretically, those approaches that have addressed organization and structure have done so within theoretical languages and models that have not been able to address the process of organizational change; they have been able to model structure and, thus, the fact of change from one structure to another, but they have not had any principled theoretical approach to the process of such organizational change. Theoretical approaches are now emerging that show promise of being applicable to such reorganizational processes (e.g., Bruner, 1976; Bickhard, 1980a; Campbell & Bickhard, 1986).

Acquiring Mental Maps as a Metaphor for Interactive Knowing

As a metaphor for considering possible organizational effects of repetitive experience, consider the task of learning to travel within a city from the initial time of confused newcomer through to the time of seasoned expert. On initial travels, one frequently lacks even the most rudimentary cognitive maps of an unfamiliar city and either scrupulously follows directions of the form "go x miles and turn right on y street; there is a Sears on the left side of this intersection," or one uses a map to provide a similar list of instructions of when to turn where. Although map usage skills could provide another context within which to illustrate the effects of experience on knowledge, for the purposes of this illustration we will not consider it and will refer to this initial state as the no-knowledge state; one gets from one location to another by following externally provided directions. However, even if you are in an unfamiliar city for a brief time for a conference, certain routes which have been traveled a few times become familiar and no longer require external instructions for their successful navigation. One might describe this as acquiring initial specific facts, and such facts lead to limited and very rigid problem solving.

As more experience is gained, certain major arteries in the city take on added importance and serve as an organizing framework for conceptualizing paths within the city and for linking components of the initial specific facts to allow the solutions to new itineraries. Thus, problem solving becomes more flexible, although it is still not optimal because the major arteries frequently do not provide the most efficient pathways. Finally, the longtime resident with the expertise of a cab driver can put together new pathways to optimally solve new itinerary problems. In addition, familiarity with the paths not chosen makes error detection more probable and error correction easier. Thus, with repeated experience an increasingly complex organization of information emerges.

Because of this emerging organization, the system of knowledge about the city is affected differently at different points in time by the experience of travelling from location A to location B. Further, several characteristics of human's mental maps make them subject to error. For example, we tend to distort reality in the direction of a right-angle grid of streets. Because of this distortion, it sometimes comes as a shock when we discover by traversing some route that two locations are much closer together than our map suggests. Under such circumstances, reorganization of knowledge is required. To complete the reorganization, repetition of prior trips is required.

As will be seen in the four research examples that follow, movement in physical space provides a good analogy for movements through other kinds of cognitive spaces. In each case, there is a cognitive space and rules for traversing that space which could be thought of as a cognitive map of the space. This cognitive map metaphor also provides an interesting perspective on the current interest with demonstrating rudimentary competence in almost everything at younger and younger ages. In the sequence just described, successful (although frequently non-optimal) navigation is apparent from the initial encounter with a new city. Although the underlying organization and use of knowledge changes dramatically, this change would not be indexed if we simply measured probability of arriving at desired destination (i.e., success rate).

Infant Number Skills

Over the past 10 years I have been involved in investigating the numerical skills of infants. As some of the initial work is described, it is important to keep in mind that for the issues of this chapter the crucial content concerns some of the developmental changes in infants' abilities to use numerical information. My work and that of one of my collaborators, Prentice Starkey, has addressed four basic issues: (1) whether infants can discriminate between visual arrays that differ only in numerosity; (2) whether infants are able to match intermodally equivalent quantities across both visual and auditory modalities; (3) whether infants possess any understanding of ordinal concepts that would allow them to make judgments with respect to relative numerosities (e.g., whether one quantity is greater or lesser than a second); and (4) whether there are any developmental trends with respect to the just-described abilities over the first year and a half of life.

Discrimination of numerosities

Are infants able to discriminate between precise small numerosities? The answer to this simple question based on a number of variants of the habituation paradigm is unequivocally affirmative. This conclusion is based on converging findings by at least three distinct research groups (Antell & Keating, 1983; Starkey & Cooper, 1980; Starkey, Spelke, & Gelman, 1983; Strauss & Curtis, 1981). These results indicate that by 2 months of age, infants are able to discriminate between arrays of two and three items (Antell & Keating, 1983) and that by 10 months they can sometimes discriminate between three and four times (Strauss & Curtis, 1981) or between four and six times (Starkey & Cooper, 1980). From the work of Antell and Keating, it is reasonable to conclude that whatever capabilities which allow infants to make these numerical discriminations are present very early in infancy. From some unpublished data which I have collected, it is reasonable to conclude that these discrimination abilities do not significantly improve between 4 and 14 months of age.

For those who are not familiar with the types of procedures used in infant habituation studies, let me briefly outline the approach. In most of my work (e.g., Starkey & Cooper, 1980), we have used an experimental method that employed duration of first fixation as the dependent variable in a standard habituation-dishabituation-of-looking procedure. In our first study 4- to 7-month-old infants were habituated to two different arrays of equal-sized dots. The two habituation arrays contained an invariant numerosity, but length and density of the arrays varied. Infants were tested in either a small-number (two versus three) discrimination or a large-number (four versus six) discrimination. In each of these number conditions, half of the infants were habituated to the small quantity and were tested with the larger, and the other half of the infants were habituated to the larger quantity and were tested with smaller. In subsequent work we have varied the size, color, and spatial arrangement of the array elements. Others have also varied the identity of the elements that make up the arrays both between and within arrays (Starkey et al, 1980; Strauss & Curtis, 1981). The variations have been used to assure that if the infants discriminate between the arrays, such discrimination must be made based on numerosity rather than on some other physical characteristic of the arrays.

The ability to discriminate based on the numerosity of arrays within a discrimination paradigm requires not only that infants detect numerical information, but also that this information is stored. Over successive habituation trials, infants must in some sense be comparing the numerosity of the present array with the numerosity of those they have seen in the recent past. Over time when the same numerosity occurs repeatedly, the infants lose interest in looking at the arrays and therefore look at them for a shorter period of time. Notice that this occurred even in the Starkey et al. (1980) study in which the identity of the common objects which made up the arrays changed on every trial. It is as if the repeated presentation of arrays with the same numerosity draws the attention of infants to that characteristic of the relationship among the arrays. After habituation has occurred, when an array of a different numerosity is presented, this deviation from the prior pattern is noticed and looking time increases. The important thing to note is that infants are detecting and retaining abstract numerical information, not particular physical characteristics of the stimuli, and that infants are using this abstract information to compare arrays.

The work on intermodal matching of numerical information (Starkey, Spelke, & Gelman, 1983) provides further data to support the claim that infants are processing abstract numerical information and are not responding to some low-level physical characteristics of the stimuli. Again let me emphasize that this is not a claim that infants understand number, but rather that they are equipped to differentiate the stimuli they encounter in ways that correspond to what adults understand as small integer set sizes. I have argued elsewhere that there is nothing in the demonstrations of numerosity perception of young infants to suggest that infants appreciate the numerical nature of what they perceive (Cooper, 1984). Infants can discriminate red from blue, squares from circles, and arrays with three objects from arrays with two objects, but the early studies of numerosity perception did not investigate any characteristics of the discrimination that would indicate what infants know about the basis of their discriminations.

Relative numerosity

Two approaches have been developed for investigating infants' understanding of the ordinal property of number. Strauss and Curtis (Curtis & Strauss, 1983; Strauss & Curtis, 1984) have used a discrimination learning paradigm in which infants are trained to respond to the greater or lesser of two simultaneously presented arrays. Transfer to arrays of differing numerosity is then assessed. Using this procedure, responding to the ordinal relationship between the two arrays has been demonstrated in 16- but not in 12-month-olds. I (e.g., Cooper, 1984) have habituated infants to the sequential presentation of two arrays which during habituation always have the same ordinal relationship to one another, e.g., the first is greater than the second. After habituation, pairs with other relationships are presented to determine if infants have detected and stored the ordinal relationship. Infants as young as 10 to 12 months remember the relationship, but 7- to 9-month-olds do not.

In large measure, these results of our more recent studies (e.g., Cooper, 1985) confirm those of our earlier work (Cooper et al, 1983), although they provide somewhat stronger evidence of relational responding in the younger infants. The 10- to 12-month-old infants clearly discriminate equality from inequality relations as the previous experiment had demonstrated, but in addition they also seem to discriminate less-than from greater-than relations. There is clear evidence to support this conclusion from the less-than condition, although the support from the greater-than condition is only marginal.

These studies of infants' understanding of number demonstrates that for young infants, it does not include the understanding of relative numerosity, although there is still some disagreement on exactly when it develops. In this paper I would like to use the model I proposed for the development of order information about numerosity perceptions as an illustration of one way in which repetitive experience can reorganize knowledge.

Explanations for acquisition of relative numerosity. Two alternative approaches may be used to account for the infant's development of the ability to respond in terms of relative numerosity relations. One is to propose a mechanism by which infants order the previously unordered numerosity perceptions. The other offered by Strauss and Curtis (1984) is to propose a mechanism by which infants become aware of or at least are able to use the information implicit in their numerosity perceptions, an accessing account. The first is the hypothesis that I have chosen to pursue: experience in detecting numerosity of groups of objects in the environment leads infants to reorganize their representation of this information as an ordered series. I have argued that the experience that is central to this reorganization of knowledge is the act of incrementing (adding to) and decrementing (taking from) sets of objects. The initial knowledge that we propose the infant has is the ability to detect numerosity and to classify transformations as either incrementing, decrementing, or other.

With these capabilities as a starting point, an infant can move in the space of small positive integers just as a new arrival can begin to move around a city, although the space in which the infant moves is obviously much more constrained. In this situation there appear to be two plausible steps that might occur early in infant number development. The first is that infants would come to realize that oneness, twoness, and threeness are all information about the same kind of thing. Using the spatial analogy again, infants discover that one, two, and three are part of the same space because moves can be made among them by additions and subtractions. The second step is that by observing the effects of additions and subtractions, infants recognize not only that oneness, twoness, and threeness are related to one another, but that they can be ordered with respect to the transformations for getting from one to the other.

Implications for small versus large numerosity. An interesting implication of these proposals is that they not only can account for the emerging abilities seen around one year of age, but they also provide a new perspective on why small numerosities are treated differently from large ones in the early preschool years when children exhibit precocity with small numerosities (e.g., Gelman & Gallistel, 1978). Young children's numerosity perception is limited to small numerosities, which leads to a number system sometimes characterized as "one, two, three, (four), many." Incrementing any of the elements in this system one through four always leads to a new element. Such is not the case for <u>many</u> which will remain <u>many</u> if incremented and may remain <u>many</u> if decremented. Hence children may learn early that small numerosities and large numerosities are different classes of things. The modification of this initial classification will require reorganization of the knowledge system later in development, perhaps with the aid of the counting system to differentiate <u>many</u> into discrete numerosities that behave just like small numerosities.

Study of infants +/- behavior with objects. Is there any evidence that infants have the kind of incrementing and decrementing experience that would make this account plausible? In some unpublished work, we have taken two approaches to answering this question. One involves a simple observational method in infants' homes. We asked the mothers of 7- to 12-month-olds to let their infants play just as they normally would. When infants played with objects, most frequently it was with a single object. When they played with more than one object, 92% of the time it was with 2 to 4 objects, and they spent a large portion of their time putting the objects into small groups and then taking objects away. Our second approach was more controlled. We placed the infant in the center of a room with 15 blue poker chips with which to play. In this case 73% of the time was spent playing with more than one and less than five chips. Again, a substantial portion of the play consisted of putting the chips together in groups and then decrementing these groups.

Summary of infant work. Thus, we have data which show that infants do not respond in terms of the numerical relation between two sets of objects until after 10 months, and we have data to show that infants have repeated experience with addition to and subtraction from small groups of objects. These data are consistent with the hypothesis that children acquire the concept of relative numerosity by experience in moving within the set of small integers by additions and subtractions. The effect of the movement within this space is to organize the space serially with additions providing movement in one direction and subtraction providing movement in the other.

Two kinds of reorganizations of numerical understanding are hypothesized to occur because of incrementing and decrementing experience. The first is discovery that the perceptions of "oneness," "twoness," and so on are detections of related information. Second, the information can be organized (at least in a pairwise manner) in terms of the concepts less than, equal to, and greater than. There is no reflective abstraction implicit in this kind of reorganization; it can all occur within a single cognitive level. Nevertheless, it represents a real reorganization of numerical understanding as a consequence of repeated experience.

Addition and Subtraction

In this section I will be discussing preschool and young school-aged children's understanding of the effects of additions and subtractions on the relative numerosity of two sets of objects. Several general claims will be made based on the studies reviewed: (1) there is a reliable developmental sequence in acquiring an understanding of addition and subtraction in this age range; (2) this sequence can be described as a sequence of rules for predicting the effects of additions and subtractions on the relative numerosity of two arrays; (3) each new rule makes use of and elaborates the previous rule; (4) in part, the development of new rules depends on discovering the limitations of current rules; (5) the development also depends on increasingly elaborate quantification of relative numerosity.

The basis for much of the work discussed in this section can be traced to a study by Brush (1978) demonstrating that some preschool children will assert that an array that has had an object added to it has more than another array even if the incremented array has many fewer objects. Thus, Brush's young preschoolers seem to be using the knowledge of the effect of incrementing and decrementing that we just discussed for infants. However, the effect of adding or subtracting on relative numerosity of two arrays is a much more complex question, and the rule young preschoolers use is correct only when the two initial arrays are equal or if the addition or subtraction increases the difference between the two arrays.

We now examine the results from several studies that assess children's growing understanding of this kind of transformation. In all of these studies, children viewed two arrays of objects constructed in temporal one-to-one correspondence. Objects were added to one of the arrays if the initial numerosities were to be unequal. Then one or more objects were added to one of the arrays, and the children were questioned about the relative numerosity of the two arrays. These experiments have provided substantial support for a four-step sequence in children's understanding of addition and subtraction: primitive, qualitative, superqualitative, and quantitative.

Primitive understanding does not differentiate "more than before" from "more than the other." The child asserts that an array added to has more, or one subtracted from has less. Thus the primitive level includes no understanding that initial relative numerosity is an important consideration in predicting the effect of a transformation. The primitive rule leads to correct predictions only when the initial arrays are equal or the transformation serves to increase their difference.

Qualitative understanding makes the differentiation between "more than before" and "more than another." The qualitative rule differs from the primitive rule because it takes into account initial relative numerosity, but does not quantify differences. One array is coded as being less than, equal to, or greater than another. An addition changes a less-than relationship to equal regardless of the initial differences. Likewise, a subtraction changes a greater-than relationship into equal. If the initial numerosities are equal, then just as for the primitive level, an addition is interpreted as making the array added to have more, and a subtraction as making the array have less.

Thus at this level, children's judgments are correct if the initial arrays are equal or differ by one, or if the transformation increases their difference.

Superqualitative understanding entails further progress in making use of relative numerosity, but still does not fully quantify it. At this level, the child differentiates between differences of one and greater than one, but fails to quantify these greater differences. In one of our experiments, a child who solved problems with the superqualitative rule labeled arrays with one more as having the "most" and arrays with more than one more as having the "mostest." The children using the superqualitative rule are limited to a five-category system for coding relative numerosity: much less than, less than, equal to, greater than, and much greater than. In general, additions of one move the child's judgments through the sequence so that, after four additions, an array that was initially judged to be much less than another would then be judged as much greater. The important addition to the qualitative rule for the superqualitative rule is the decision that differentiates between arrays that differ by one and arrays that differ by more than one. An exception to the general rule occurs if the arrays differ greatly in numerosity. If, after an addition, the array added to is still clearly much less than the other, the second addition will not lead to a judgment of equality; this is an example of a quantifier rather than operator solution. Despite this more sensible integration of quantifier information at the superqualitative level, it is still the case that the child's understanding of addition and subtraction provides correct "operator" solutions only when the initial arrays are equal or differ by one or two.

Finally, *quantitative* understanding involves quantifying the initial differences between arrays and quantifying the amount added (for example, by counting on solutions) to arrive at fully accurate performance. It differs from the superqualitative rule in the type of quantification used for relative numerosity.

Table 1 summarizes the results from five studies designed to test the level of understanding of addition and subtraction in children from 2 to 7 years of age (Cooper, 1984). The percentages do not add up to 100% for all age groups because some children could not be classified at a single level. In these studies, primitive understanding predominated for 2-year-olds (2:0 to 2:11); qualitative understanding, for 3- and 4-year-olds; and quantitative understanding, for 5-year-olds and older. Some of our more recent work suggests that these children may have been precocious by about 6 months compared to the average population in Texas. Note that in all age groups, the percentage of children detected in the superqualitative level is very low; 13.2% for 6-year-olds is the highest level observed.

Table 1. Percentage of Children at Each Level of Addition/Subtraction Understanding for Large Numerosities (n>5).

Age	Primitive	Qualitative	Super-qualitative	Quantitative
2	62.5	12.5		
3	22.2	55.5	11.1	0.0
4	36.6	56.7	0.0	0.0
5	15.0	25.0	10.0	50.0
6	17.0	13.2	13.2	56.6
7	7.2	14.2	7.2	71.4

The small number of children who are classified at the superqualitative level led to two questions. Does superqualitative understanding really exist as a separate approach, or is it an epiphenomenon from errors of children at the qualitative and quantitative level, and if it does exist, do all children go through this level of understanding in their progress toward quantitative understanding? Two longitudinal studies have been conducted in which children were selected for being at the qualitative level of understanding and then were tested repeatedly. In the first (Cooper, 1984), 24 children were tested on the average of every 3 weeks for a period of 7 months. At the end of the study, 17 were classified as quantitative and 2 as superqualitative. Of the 17 who became quantitative, 10 were never classified as superqualitative. In the second study (Cooper, 1987a), 36 children were studied with 4 assessments in a 3-month period. In the course of the study, 4 children became and remained superqualitative and 9 became quantitative. Of this later group, 4 were observed at the superqualitative level for at least 1 testing session. From these two studies, it is reasonable to argue that the superqualitative approach becomes a stable enough approach to these kinds of problems that it can be detected in about 50% of children using a monthly assessment. One might argue that the repeated testing involved in these studies sped development and the superqualitative approach may have greater longevity in the population at large.

In the second longitudinal study just described (Cooper, 1987a), three other tasks were also administered: conservation, inference in which children had to say what transformation had occurred by viewing two arrays before and after some transformation, and the transfer task described earlier. Improvement on these tasks was correlated with improvement on the standard addition/subtraction task. Further, children who showed quantitative understanding without going through the superqualitative level showed the greatest improvement on the other tasks. Finally, when teachers were asked to rate children's number skills as poor, fair, good, or excellent, children who were detected at the superqualitative level were less likely to be rated as excellent than other children in the study (even those who remained at the qualitative level).

Summary of addition and subtraction studies. The studies described here document a reliable sequence of 3 levels which characterizes children's developing understanding of the effects of additions and subtractions on relative numerosity, and a fourth level of understanding which some children go through and all children may go through. Just as development in infancy was described as learning about movement in the space of small numerosities as a function of additions and subtractions, the development described in this section can be conceived of as learning about movement in the space of relative numerosity as a function of additions and subtractions. This learning involves detecting the relationship between a start point (initial relative numerosity), a transformation (addition or subtraction), and an end point (final relative numerosity). Each developmental step leads to a system which makes fewer errors than the previous level, although it is not until the quantitative level is achieved that errors are eliminated. Error detection can provide the feedback that something needs to be changed, but it does not specify what.

In order to consider what changes, let us re-examine the four levels. For the primitive level, initial numerosity is ignored, and only the transformation is considered in predicting final relative numerosity. For the qualitative level, initial numerosity is considered, but only with a three-state qualitative description: array A is less-than, equal to, or greater than B. For the primitive level, the predictions are consistently correct only when the initial numerosities are equal. For the qualitative level they are consistently correct when the initial numerosities are equal or differ by 1. For the superqualitative level, they are consistently correct when the initial numerosities are equal or differ by one or two. Finally, for the quantitative level, they are consistently correct for all initial numerosities. Notice that part of the change that this progression highlights is an increasing differentiation in how the initial numerosities are represented.

When recently describing these results, I was challenged to explain why children were progressively acquiring solution systems that were wrong. One form of answer to this question is to make an analogy to Newtonian physics and ask why Newton made the errors inherent in classical physics. Within the realm of data for which he was seeking explanations, Newton's approach worked quite well. For young children, their primary mode of quantification involves their perceptual skills for detecting numerosity. Since these exist in a very limited numerosity range, the opportunity to detect the errors in the qualitative or superqualitative levels is limited. Additionally, even when it becomes clear that a given approach does not always lead to the correct answer, if it is correct some of the time, it will not be abandoned until a better approach is found.

As children learn to count, counting provides important information about whether judgments of relative numerosity are accurate, but it does not directly lead to additional solutions. However, repeated experience using counting or one-to-one correspondence not only provides information about errors, it also provides information for reorganizing information about relative numerosity, e.g., quantifying how much more. The superqualitative level is particularly important here because it documents that a concern with the question "How much more?" precedes full quantitative understanding (at least for some children).

Preschoolers' Understanding of One-to-One Correspondence

At least since the time of Piaget's first studies of number conservation, judgments of the relative numerosity of two sets of objects frequently have been used to study preschool children's understanding of number concepts. In a study of children's understanding of the effects of addition and subtractions on relative numerosity (Cooper, Campbell, & Blevins, 1983), we serendipitously discovered that a "growth error" frequently occurs around 5 to 6 years of age, in which children will no longer accept as evidence for equality temporal one-to-one correspondence in the construction of two sets. This observation led to a more systematic cross-sectional assessment of the kinds of information that children would use in judging equality, followed by a short-term intervention designed to provide experience to the children and one which would be relevant to information selection and use.

In this study, eighteen 4- to 5-year-olds, eighteen 5-1/2- to 6-1/2-year-olds, and eighteen 7- to 8-year-olds were asked to judge the relative numerosity of two arrays under a variety of conditions. Small (2 to 4 items) and large (7 to 11 items) numerosity arrays were used, but for our purposes here we will discuss only the large numerosity results. In the *spatial one-to-one correspondence task*, the items in the two arrays were placed in rows one above the other with equal spacing so that one-to-one correspondence could be used to judge equality or inequality. If the arrays were unequal, an internal item was missing so that one-to-one correspondence but not length information could be used to make the correct judgment. In the *simultaneous temporal one-to-one correspondence task*, the two arrays were constructed so that each time an item was added to one array, an item was simultaneously added to the other array. For the unequal trials in this task, an extra item was initially placed in one of the arrays for half of the trials, and an extra item was added at the end of construction for the other half of the trials. As the two arrays were constructed, the items were placed in two heaps to avoid obvious perceptual information about relative numerosity. In the *successive temporal one-to-one correspondence task*, items were alternately placed in one and then the other array; as before, the items were placed in two heaps.

Results of Pretest

Four-year-olds were successful on both the spatial and simultaneous temporal one-to-one correspondence tasks but did not perform at an above-chance level on the successive task. Six-year-olds were successful on the spatial task but frequently argued that one would need to count to "know for sure" on both the simultaneous and successive temporal tasks. When pressed to make their best guess, they were correct at above-chance levels on both tasks (74% and 67% correct) but performed more poorly than the 4-year-olds on the simultaneous task (93% correct). Seven-year-olds performed extremely well on all three tasks.

Training

After the initial assessment, the two younger groups were given two "training" sessions which consisted of repeated experience with either the simultaneous or successive construction procedure, with nine children from each age group experiencing each training type. Two arrays were constructed as in the initial assessment. The children were asked to make a relative numerosity judgment, then to count the two arrays, and then to again make a relative numerosity judgment. Finally, if they judged them to be unequal, they were asked to make them equal by putting on additional items or taking items away.

Results of Posttest

Following the two training sessions, a final assessment session was conducted which contained tasks identical to the initial assessment session. In addition, new versions of both the simultaneous and successive temporal one-to-one correspondence tasks were included in which additions to the two arrays were made in groups of two or three items (consistent within a trial and varying between trials). Neither training significantly affected the performance of the 4-year-olds. Note that the children were already at ceiling on the spatial and the simultaneous temporal task, so one can conclude only that this training did not facilitate performance on the sequential task. The modification of the task, which involved the addition of two or three (rather than one) items at a time, led to more errors on the simultaneous task, although performance was still well above chance (83%), and performance on the successive task remained at chance.

The pattern of performance for the 6-year-olds was very different. They showed improvement on both the simultaneous and successive tasks regardless of training group, and surprisingly there was no difference between the two training groups (96% on the simultaneous task and 79% on the successive task). This improvement was also apparent on the tasks when additions of more than one were employed (97% on the simultaneous task and 76% on the successive task).

Discussion

All children in these studies were already aware that adding to an array increased its numerosity and subtracting decreased it (e.g., Cooper, 1984); it is their understanding about the effect of these operations on relative numerosity whose development is being examined in this study. Just as on conservation tasks, 4-year-olds will accept a variety of different kinds of perceptual information as indicating equality or inequality. The conservatism of the 6-year-olds is, in fact, correlated with improved performance on conservation tasks. These data are consistent with the hypothesis that the acquisition of conservation leads to a reorganization of the criteria for equality and inequality. One might say that the discovery that past procedures have been in error makes the 6-year-olds very cautious and in need of further exploration in the space of equal and unequal relative numerosity.

What is particularly impressive in this study is that the re-exploration of the equality and inequality space allows the 6-year-olds to make use of the similarity between the simultaneous and successive tasks, a similarity to which the 4-year-olds appear to be oblivious. Although I cannot detail the argument here, I would suggest that it is the intermediate move to using counting as the sole criterion that facilitates this discovery. From an interactive perspective, there is a reorganization that occurs between 4 and 6 years. The role of training in this study was to leave the organization of knowledge of the 4-year-olds unchanged and to expand the newly reorganized system of the 6-year-olds.

Children's Acquisition of Some Algebraic Manipulations

The fourth example comes from our current work on the acquisition of algebra skills by elementary school students. The data to be reported come from a more extensive study of the acquisition of algebraic concepts of second through fifth grade students who are participating in a special weekly mathematics training program. One goal of the program is to adapt Suzuki's method of teaching violin to the teaching of mathematics in the elementary school years. The two principles of that method that are relevant to this paper are the breaking of tasks into components on which learners have a high probability of success and the use of repetition as a technique for perfecting the subskills and integrating them into larger units.

Training procedures

In this paper, I will use children's performance on problems of the form $x - 3 = 10$ and $x + 2 = 7$. As an aside, I might mention that training elementary school children to translate word problems into algebraic expressions is not the impossible task that some might imagine. However, what is of interest here is how children learned to solve these very easy algebraic expressions for the value of x. Let me use the $x - 3 = 10$ expression to illustrate the teaching approach in a problem where x represents the age of a child, John. I will give only the teacher's side of the conversation, but the child's responses can be easily imagined.

> "So, three years less than John's age is 10."
>
> "If 10 is three years less than John's age, what would be two years less?"
>
> "Great, and if 12 is one year less than John's age, how old is John?"
>
> "All right, let's see what we've done. We started with two things that were equal: $x - 3$ and 10. Then we added 1 to each of them. If we start with something that is 3 less than x and add 1 to it, we get something that is 2 less than x or $x - 2$, and if we add one to 10 we get 11".

The training continued in this form, emphasizing that if you add 1 to each of two things that are equal, they remain equal. Notice that this is exactly the same training that occurred in the previous study of preschool children's understanding of one-to-one correspondence, except that the incrementing there was done with real objects.

Three Training Groups

One group of children was taught this successive incrementing-by-1 technique to solve many problems of the form $x - m = n$. Another group of children was taught a successive decrementing-by-1 technique for solving problems of the form $x + n = m$. A third group was taught both techniques. The children were then given many problems to solve. Since translating the word problems into algebraic expressions is initially quite challenging for 7- to 11-year-old children, the days of practice did not constitute boring repetition.

Shortcut of Add/Sub and Generalization to Other Operator

As the children solved more problems, their performance revealed two changes from the procedures explicitly taught that are interesting from the point of view of this paper. First, 94% of the time, children collapsed the process of adding or subtracting one n times into adding or subtracting n in a single operation. Second, for those children who had been taught to solve problems only by adding or only by subtracting, 63% immediately solved the other kind of problem the first time they encountered it after three days of their specialized training. The first case, the collapsing the incrementing or decrementing-by-1 n times, offers an example of change with practice which could be accounted for within the kind of model that involves the combining of production rules. However, in the second case, the generalizing knowledge about how to solve problems of the form $x + m = n$ to problems of the form $x - m = n$, a much more complex effect of experience is involved.

Discussion of Generalization to Other Operator

Notice that the generalization from solving problems by incrementing to solving problems by decrementing does not involve applying a procedure used in one case to solve another. Rather it is knowledge or understanding of the problem that is generalized, and then a new and appropriate procedure is invented. One way to describe this process is to note that children of this age have already learned to move by addition and subtraction through the space of the positive integers. In our training, we then provided them with the opportunity to learn how to move in the variable space of a simple algebraic expression. The generalization involved here involves reflecting on the structure of the two spaces using the understanding they have gained about the positive integer space as an organizing principle for understanding the algebraic space. Note additionally that it was training in moves within the algebraic space that provided the basis for this generalization.

Summary and Conclusions

In the four cases I have presented from children's number development, a number of common threads are apparent and are indicative of the specific interactive model for number development that I have been developing over the past several years. In all four cases, the *focus is on addition and subtraction* as operations which are central to processes of generating, elaborating, or reorganizing numerical spaces. In the infancy work, the points in the space are *specific small numerosities* which become organized in a serial order with respect to the operations of addition and subtraction. In the preschool data, it is a *relative numerosity* space where again the operations which define it are addition and subtraction. With the elementary school-age children it is *a variable space*, and again the operations which define it are addition and subtraction.

Experience as Generating and Modifying Cognitive Space

The role of experience in each case is to generate or modify the organization of the relevant cognitive space, just as in the cognitive map example the role of experience was to generate or modify the cognitive map. It is important to note that at one level it is the constructed cognitive space that constitutes knowledge, and it is for this reason that repetitive experience is important. Let me clarify by examining an alternative. Suppose that cognitive development proceeded by a set of genetically provided inference rules. In such a case, one example (or a few examples depending on the nature of the rules) would suffice to support the inference. On the other hand, if it is the space generated by successive interactions that constitutes knowledge, then repetitive activity is necessary to fill out the space. Just as in the case of the cognitive map example, the space does not have to be fully saturated to provide useful knowledge. Nevertheless, as the space becomes more richly interconnected, a wider variety of moves within the space is possible.

Reflective Abstraction

The role of repetitive experience generating richly interconnected spaces is also important for reflective abstraction. Reflective abstraction occurs with respect to the structure of cognitive spaces, as was illustrated in the algebraic example. The structure of the spaces must be sufficiently well defined so that the interrelation among spaces can be detected.

The role of repetitive experience that has been emphasized in all four examples in this paper is as a provider of information that facilitates problem solving. Hence, I have not emphasized the repeating of solution strategies so that they become effecient or automatic (the effect that the term practice cannotes). I have not emphasized the development of prerequisite skills that then become components in later, more complex problem solving systems. These may also be important components for cognitive growth, but the role of experience on which this paper focuses is both central to cognitive development and often neglected. Mathematics learning provides an ideal environment in which to study the role of repetitive experience in reorganizing systems of understanding. In addition, understanding its role will help in curricular planning with respect to such issues as the role of drill and practice in mathematics education.

8 The Concept of Exponential Functions: A Student's Perspective

Jere Confrey

Exponential functions represent an exceedingly rich and varied landscape for examining ways in which students construct their understandings of mathematical concepts. They are useful in modelling phenomena across many fields, including astronomy, economics, chemistry, biology and information theory, and are typically the mathematical "tool of choice" for exploring "bigness" and "smallness," bringing these concepts into the range of human comprehension. They provide a dramatic contrast to linear functions in that the change they model is based on repeated multiplication instead of repeated addition; thus their rate of growth can be shocking.

Since exponential functions can be useful in eliciting students' conceptions, they provide an ideal "critical research site" (Shulman, 1978). This site is enriched by the isomorphic mapping between addition of exponents and multiplication of exponential expressions. It creates a system rich in interconnections in which one can explore students' beliefs about the logical consistency across the mapping.

The work described in this paper is conducted in the constructivist tradition of extended teaching experiments, which have typically involved students in the elementary grades (Cobb and Steffe, 1983; Steffe, 1991; Steffe, von Glaserfeld, Richards and Cobb, 1983). One aim of this research is to extend this approach beyond the elementary grades. This extension will involve an examination of how the teaching experiment changes when used to explore students' conceptions of more complex topics, particularly when those students have experienced years of traditional instruction. Such investigations can offer insights and techniques to the researcher engaged with younger children.

Unlike earlier work documenting students' misconceptions (Confrey, 1986; 1990), the intent of this work is to describe how students construct an understanding of exponential functions and related concepts in a flexible, ideographic, and functional way. As we develop models of the various paths our students take, we compare them with the traditional presentation in mathematical texts to see how the socially defined, mathematical conventions expressed in the text resonate with or deviate from student thinking. In the course of doing so, our own understanding of the mathematics is transformed. We will suggest that the documentation of the change in our own mathematical understandings constitutes as legitimate a research outcome as that of the students.

In this paper then, I will begin with a brief portrayal of the treatment of exponential functions in traditional materials. An analysis of the form of mathematical argument embedded in the portrayal will be offered and connected to some of the assumptions in the formalist view of mathematics. I will then discuss an alternative theoretical perspective on what mathematics is, drawing from constructivism and the philosophy of science. Based within that theoretical framework, I will discuss one case of a nine-hour teaching interview on exponential functions. Following a discussion of methodology, a model of the student's developmental route will be offered. The paper will conclude with a discussion of the implications of the theory, methods, and results for the practice of educating in mathematics.[1,2]

The Traditional Account of Exponential Expressions and Exponential Functions

The treatment of simple exponential expressions and functions in precalculus texts in the United States is relatively uniform. For the purpose of this paper, we will restrict our description of the traditional account to precalculus level topics; future work will explore the more advanced topics. We use the term "exponential expression" to refer to algebraic or numerical expressions in the form a^b where b is numeric (i.e., 2^3, 5^{-3}, x^3, $x^{1/2}$). Also, exponential functions are those able to be codified in the form $f(x) = ca^x$ with $a > 0$ but $a \neq 1$ and c equal to a numerical constant.

Most books treat exponential expressions by introducing positive exponents as abbreviations for repeated multiplication. This accessible, but ultimately incomplete (when extended to the reals), definition of an exponent as a *counter* is used to justify the rules (sometimes referred to as properties) of exponents. These rules are defined with the following constraint: the exponents must be positive integers. Thus, when dividing exponential expressions, the magnitude of the exponent in the numerator must exceed that of the denominator.

Most texts then offer a "plausibility" argument, suggesting that if one wanted to create a meaning for a negative, zero, or fractional exponent, it should be done to preserve the consistency in the rules (x^a/x^a should equal $x^{a-a} = x^0$ and, since $x^a/x^a = 1$, then $x^0 = 1$). These plausibility arguments are then legitimated by postulating definitions for x^0, x^{-a}, and $x^{1/a}$ which thereby allow one to relax the previous constraints on the rules.

Thus the rules, which originally derived their meaning from the definition of an exponent as an abbreviation for repeated multiplication, become the grounds on which the new definitions are based. Those new definitions no longer make sense as an example of the initial definition (to equate "x^{-3}" with "x multiplied times itself negative three times" is nonsensical). This extension of the domain to all rational numbers is achieved by sacrificing the intuitive appeal of the initial meaning and gaining the elegance of a broad isomorphic relationship between the exponents and the exponential expressions.

Once the rules are extended, students are typically expected to engage in the completion of a set of exercises which require the faultless application of the rules to complex expressions. The few applications offered in these sections are primarily oriented towards the use of scientific notation.

Exponential functions are usually introduced quite quickly as a class of functions of the form $f(x) = ca^x$ with $a > 0$ and $a \neq 1$. As with other functions, the stress is on their domain and range and on obtaining a correct graph. Seldom is the question of the continuity of the graph and the meaning of irrational exponents even mentioned. Some attention is given to the shared characteristics of all the graphs, pointing out that, in the form $y = a^x$, they all pass through the point (0, 1) and (1, a) and are asymptotic to the x-axis.

A quick mention of the function $f(x) = \log_a x$ is used to introduce students to logarithms. Typically, this is followed by a lengthy discussion defining the expression $a = \log_b c$ as equivalent to $b^a = c$ and deriving the rules for logarithms from those for exponents. A change of base formula may be provided. The exercises then stress ways either to switch from logarithmic notation to exponential notation and vice versa or to simplify logarithmic expressions.

Logarithmic functions (base ten) are then developed as inverses of exponential functions which have the properties of logarithms. Some practice with applications are provided such as with pH, Richter scale, etc. These applications are given as formulae which are to be solved for a single unknown. The introduction of e is most likely to be encountered within a cursory discussion of the natural log, and it is presented as an alternative base and is given a name. Possibly a definition such as $e = \lim(1+ 1/n)^n$ is provided, but its discussion is limited to avoid appeals to calculus-based arguments.

The organization and presentation of the typical curriculum lends itself to the following analysis:

1. The material is offered in a logical sequence which "covers" the topics but pays no attention to the psychological issues such as the conflict created for students when the extension of the exponent to non-positive integers prevents them from using their accessible meaning as an abbreviation for repeated multiplication.
2. Applications are cast as circumstances in which to demonstrate the usefulness of the techniques, but in no instance did we find an application presented in such a way as to encourage the student to consider why an exponential function was the appropriate choice of a model. Thus they are viewed as formulae rather than as opportunities for relating the material to everyday experience.
3. Students are required to demonstrate facility with the manipulation of expressions above all else. Their engagement with extensions of the mathematical structures or with applications is minimal.
4. The argument structure in the text is: definition, rules, plausibility by extension argument, definition, proof, and elaborated exercise. The arguments to extend the use of exponents beyond positive integer values may conceal what is actually a complex and difficult set of issues concerning the isomorphism between structures involving the exponents and exponential expressions. When students

are limited to the positive integral exponents, the "meaning" behind the exponential expression is accessible and preserved. When efforts are made to expand the isomorphism to the set of all integers, all rationals, or all reals, a *fundamental shift* must occur in the *meaning* of the exponent. The isomorphism becomes the basis for meaning rather than the view of the exponent as a counter.

5. Finally, the traditional presentation of exponential functions minimizes the underlying multiplicative operation. This failure to emphasize the operational character of functions increases the likelihood that students will fail to recognize the call for a particular function in a contextual situation, and it further increases the likelihood that concepts such as doubling time or half-life will not be understood in regard to the implicit constancy of multiplicative rate.

In summary, although one does not want to presume that a text serves as a complete representation of an instructional program, one might at least assume that it is indicative of the path one expects students to traverse in their acquisition of the material. The texts reviewed demonstrate that the presentations are dominated by rather a formal but tenuous series of definitions and rules followed by extensive practice in symbol manipulations. Rules often mask the broader systematic qualities of the relations, and the underlying operations of the function are lost in the presentation of the general form. Applications are not used to motivate nor to connect to the experience of the student. Instead they are treated as formulae and opportunities for more manipulation of symbols.

The purpose of this analysis is to emphasize that textbooks are based on formal mathematical argument and that, when one examines the development of a mathematical idea with students through the use of constructivist teaching experiments, it becomes obvious that a formal presentation is simply inadequate to promote insightful learning. However, to offer an alternative, one must study the genesis and evolution of a mathematical idea, place the mathematical content in contextual problems which highlight its mathematical function, and document the pathway students traverse in gaining insight into the idea.

Method

The following section contains some observations from the analysis of a series of nine one-hour teaching interviews with one student, which were conducted as a part of a larger project on students' understanding of exponential functions. The overall study includes a set of six such intensive case studies, a review of the historical development of exponential functions, an examination of the cross-disciplinary use of the concepts, an analysis of full-class data from a precalculus course, and the development of visual, multiple-representational contexts for examining students' conceptions.

The student discussed in this case study is an animal science major at Cornell University. At the time of the study, he was enrolled in a remedial algebra course at the university. His SAT math score was 540 and his verbal score was 530. He volunteered to participate in the study and was paid for his participation. Overall, as the interviewer, I would describe him as bright, very clever in imagining potential relationships, insightful in his self-assessment of his understanding, poorly educated in understanding mathematics, and weak in his ability to monitor his local goals. He was relatively transparent in terms of signalling his confusion or satisfaction and, so to speak, wore his heart on his sleeve.

Individual interviews of the teaching experiment were conducted twice per week over four and a half weeks. Each interview was videotaped and transcribed. The problems for each interview were predesigned but subject to modification over the course of the experiment. After the first interview, the student was asked to take home and complete a pretest covering the skills in manipulating algebraic and complex exponential expressions and in graphing functions typically covered in the secondary curriculum. The purpose was to provide a baseline that would allow a comparison between "school" skills and issues that may arise in his solutions to problems.

After discussing the pretest, he was asked to draw a map of some planets whose distances from each other and the sun were given in scientific notation. These were followed by problems on depreciation and exponential growth, linear erosion, binary search, and biological growth. The last interview was a variation on the first, having a sorting task combined with a discussion of the overall teaching interview.

It was presumed that the student always made sense within his conceptual framework and the role of the interviewer was to investigate that conceptual framework through the use of flexible probes. Each interview began with a request for the student to discuss the previous session. After each interview, the interviewer and the camera operator discussed the session, and the interviewer wrote a set of notes on the session. Between sessions, the tapes were viewed (without transcription) and discussed and a plan was developed and written for the next interview.

The videotapes and transcripts have been analyzed by the interviewer and a team of viewers consisting of graduate students and staff in mathematics education. These analyses often took the form of editing a set of transcripts with commentaries. The commentaries led to the accumulation of recurring themes which seemed to possess strong explanatory value. A story line of the student's progression through the sessions was proposed and critiqued. Certain tape segments were identified as being especially significant and representative of the student's approaches and development and were viewed and discussed repeatedly.

The interviews have been labelled "teaching experiments" because it was expected that the student would go beyond his previous familiarity and hence *learn* over the course of the interview (Opper, 1977; Steffe, 1991). That students experience learning during interviews has been documented even in short interviews and is an important quality of the work (Confrey, 1981). All too often, it is ignored or an apology for it is offered because it is evidence that the interview is indeed an intervention. In this work, the role of the interaction between interviewer and interviewee is acknowledged as an intervention; it is assumed that, when students engage with a genuine problematic, struggle to resolve it, and reflect on those actions, learning of a very profound sort has occurred.

This approach also recognizes that learning in these sessions is not haphazard but occurs within a socially constrained conversation between someone who is more knowledgeable and someone who is less knowledgeable, a fact known to both participants (Mischler, 1986; Brousseau, 1984a). It is also the case that the interviewer does not presume to know beforehand how a student will proceed at any time or what the student knows. Thus the interviewer must be prepared by her/his experience with this concept to anticipate and seek out with sensitivity the variety of approaches tried by the students. However, as interviewer, I often found the student's diverse methods surprisingly imaginative; and at times, on reviewing the tapes together with the research team, I came to believe that I had pursued a direction that neglected or was in contrast with what the student meant or intended.

The goal of the retrospective analysis is to explain how a particular student has built up his understanding of exponential expressions and functions. Interpretative frameworks are built through an analysis of characteristic expressions and actions engaged in by the student while solving problems. These characteristic ways of acting, which appear fundamental in their usefulness and recurrence, are subject to intensive examinations. Other statements, which occur only once or which do not make sense to the interviewer are passed over. It must be stressed that this is interpretative work; the interviewer pursues only the events which are at least vaguely recognizable to her, just as the analysts can make sense of the data only within their own limited frameworks.

Through the process of the interview, my own conception of exponential functions was transformed, elucidated, and enriched. Our minds are such sensitive instruments, and as we strive to examine our understandings, the character of those understandings will inevitably shift; this is the essence of the process of reflection and construction. In order to view the mathematics from the perspective of a student, interviewers accept the constraints that the student puts on our mathematical knowledge. Functioning within these constraints invariably leads the interviewers toward new insights in the mathematics--and they find out what they are most committed to and what they are willing to sacrifice and change. Our own self-understanding becomes particularly keen when our goal is to not only describe a student's capability and current beliefs, but to model the process of change itself as it unfolds in both form and content.

Our methodologies must allow and assist us in speaking of those transformations in our own understandings and how these transformations were mirrored, inspired, and woven by our examinations of the pursuits of others. The transformations of the research team may provide a significant resource for teacher educators seeking to assist practicing teachers to develop pedagogical expertise.

It is hoped that over the course of examining a variety of students' work and of conducting cross-disciplinary and historical work, clearer discriminations can be offered between the researchers' and students' transformations. However, early on in such research, the goal is to create rich contexts. David Hawkins (1974a), in an essay entitled "I, Thou and It," speaks of the importance of creating contexts imbued with bounded but rich and novel subject matter for children to explore in the company of adults. He emphasized that the subject matter in such a situation is not simply an "other" added to the triangle of I and Thou, but it is a mirror and a vehicle for coming to know oneself and others. Such a portrayal suggests that the separation of I, Thou, and It can never be achieved absolutely or exhaustively, but only in degree. To pretend to be able to do otherwise is to ignore or underestimate the participatory role of observer, the vitality of concepts, and the sensitivity of the students with whom we work.

Results

These results are offered from the perspective that mathematics consists of mental operations that are useful to human beings in organizing and acting in their environment. It includes the records of those mental operations and our ways of communicating about them. According to this perspective, no knowledge exists which transcends human experience. Thus, to explain the construction of mathematical ideas, we need to look at how our characteristic ways of viewing, acting, and operating in the world change and at how this change is communicated to others. This I take to be the core commitment of the constructivist researcher in mathematics education (Confrey, 1985; von Glaserfeld, 1982).

In this view, one cannot easily separate the psychological and the epistemological. For the constructivist, the individual constructs his/her understanding of the world within a social framework which exerts a profound influence on the course of those constructions. An examination by experts in mathematics education of an individual striving to make sense of mathematical content has legitimate epistemological content for the constructivist. This is not to say that such content is evident to the interviewee, but that it can be discerned by experts and is useful in explaining the interviewees' actions.

Thus, the discussions which follow are not only a psychological portrayal of an individual student, they are part of an epistemological attempt to provide a portrayal of the meaning of the concept of exponential function as it is constructed by humans. This is done through a set of five interpretative frameworks which have been useful in modelling this student's understanding

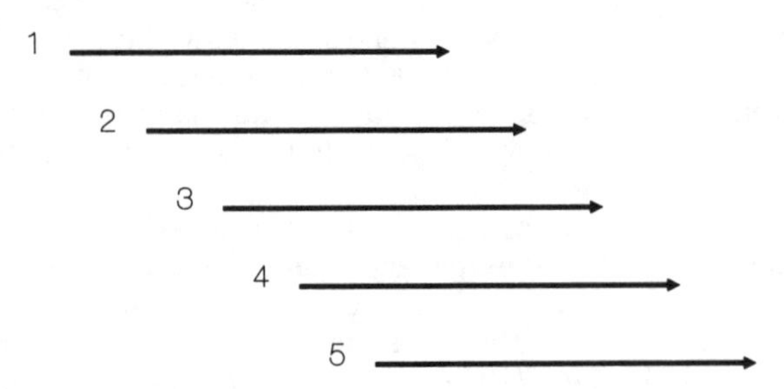

Figure 1. Relationships Among Frameworks

of exponents, exponential expressions, and exponential functions in the interview setting.

These five interpretative frameworks are not intended as linear developmental stages. In complex exchanges, one may see evidence of all five. They are roughly ordered from the simplest insights to the more complex, but they should be viewed as overlapping. Perhaps an image such as that represented in Figure 1 would best describe their relationship.

Framework One: Exponents and Exponential Expressions as Numbers.

In the traditional presentation, negative and fractional exponents are seen as simple extensions on the class of positive exponents. It should be clear to any experienced teacher that they are not simple extensions for students. As Dan said, "And then I figured 9^{-2}, how could you do nine times itself negative two times?" Later, in the case of a fractional exponent ($9^{1/2}$), he said," It's just like, kind of times like, like 9^2. You know, you can't times nine by 1/2."

9^0 is no less troubling to Dan. In the case of 9^1 and 9^0, when asked to write out the values for 9^3, 9^2, 9^1, 9^0, he comments, "I'd just put nine times nothing [for 9^1]. So my reaction [here for 9^0] would be just to put, just to put nothing down as the answer. If there is not noth-, put nothing down."

Later, when exploring 9^0 as related to $9^5 / 9^5$, he comments, "When I see this problem [$9^5 / 9^5$], I don't think at all of this [9^0]."

Thus one can suggest that for Dan, initially, *exponential expressions are abbreviations* for counting and recording the number of repeated multiplications and thus almost an ideographic representation. The accessibility of this definition facilitates the connection between the representation and the action it entails, but this connection fails when one attempts to extend the meaning into the case of zero, negative, or fractional exponents.

When moving into the arena of fractional and negative exponents, students expect to find that negative exponents produce negative numbers rather than fractions and that fractional exponents produce fractions rather than radicals. Although they may learn correct procedures for manipulating expressions with fractional or negative exponents, these expectations are never directly addressed, and they resurface when students are faced with non-routine tasks.

Dan exhibited repeated confusion when dealing with fractional or negative exponents as demonstrated when he was asked to place 1/3, 9^3, 9^{-3}, 2, 9^0, 10, -3 and -1 on the number line. First he assumes that a^{-x} should be as equally distant from zero as a^x, and in doing so, he draws from his understanding of the relationship of x and -x on the number line. When asked to place 9^{-2}, he seems to recall that negative exponents are related to fractions, and he thereby transforms it into $9^{1/2}$ and claims it is equal to 3:

> And 9^3 is somewhere over here (points out towards right edge of the piece of paper) and 9^{-3} is somewhere exactly in the opposite direction, over here. And I probably should have just put these in.

> 9^{-2}: well, it's nine times 1/2, which is three. I know that's three. Uh, alright, well, I'll just put it in three. I don't know why, but I know it's, I know it's right here [at 3].

Methodologically, Dan's discrepant performance on 9^{-2} and 9^{-3} raises interesting questions. If, as interviewers, we believe that Dan is always making sense within his own conceptual framework, we surely must base our definition of this ideographic sense-making on our perception of consistency in students' methods. Yet, examples such as this one, of rapid and proximal inconsistency are not uncommon. One explanation would be to deny that Dan has "learned the rules," claiming that he lacks a certain competence. However, to do so would be to ignore the fundamental challenge to our assumptions that people act in rule-governed ways raised by this instance.

A second, perhaps more attractive, alternative would be to explain Dan's performance as an example of a student with competing frameworks, where the framework for negative numbers as symmetric around zero competes with the framework for recognizing exponents as square roots. This type of analysis, similar to the claim of DiSessa (1985) concerning "knowledge in pieces" and Davis' (1980) alternative frameworks, offers us a way to interpret such apparently patchwork behavior patterns and to view the student as a decision-maker at some level of awareness. With no further probing by the interviewer, we cannot specify any decision criteria. One possibility could be that, in the case of 9^3 and 9^{-3}, Dan focuses on the symmetry of the pairing of numbers ± 3, whereas in the other case of 9^{-2}, he focuses on calculating $\sqrt{9}$ and on placing that value. In this, one witnesses the close connection between the action which spawns a number and how that number is viewed as a "number." That is, there are many concepts of number connected to the variety of actions that produce them.

One way to interpret this episode is to suggest that what we are witnessing, as Dan tries to make sense of negative and fractional exponents, is his attempt to *locate the boundaries of the initial definition.* Thus we suggest that when he encounters these anomalies to his present primitive (but functional) concept of exponents as abbreviations, he must reconstruct the boundaries of the original concept and then carefully extend them into the new setting. In the studies we've completed, it becomes clear that rule-governed behavior (wherein a student identifies an instance as a member of a class and acts according to a rule) is simply not a rich enough framework to explain Dan's behavior. The concept of competing frameworks is somewhat more adequate as a form of explanation, but it seems to underestimate the purposefulness in the student's action. It lacks the capacity to allow us to tell the story of the genesis of a consistent framework gradually over time and over specific kinds of examples.

We suspect that this may be due to its lack of sufficient attention to students' perception of the tasks provided by the interviewer, what I have termed the student's problematic (Confrey, 1986; 1991), and to what they believe constitutes successful completion of those tasks.

When Dan is asked to place points corresponding to exponents on a number line, he is faced with a relatively atypical task. Thus expectations (like the belief that, since +a and -a are symmetric around zero, so too are expressions of the form a^{+m} and a^{-m}) surface which may have been tacitly held, were functional in past circumstances, and become problematic only in this novel instance. The likelihood of witnessing largely unintegrated actions is increased, and perhaps any expectation on the part of the interviewer for the student to have immediate awareness and resolution is unreasonable.

Perhaps more satisfactorily, we as researchers need to develop a progressive explanatory model which allows more variable, locally inconsistent, but increasingly coherent actions to explain a student's path to understanding a concept. Such a model would be developed from significant moments, portrayals of the student's problematics, descriptions of his/her choices of actions, and perhaps movement towards his/her establishment of a systematic and robust way of acting on the material. The model would be based on the assumption that it is not the performance of routines that is interesting, but the act of conceptualizing the problematic and then creating or selecting ways to resolve it.

The Impact of Ambiguous Language. The language of exponents as spoken English seems to contribute to Dan's challenges for integration and to provide evidence of sources of confusion. In Dan's tapes, his lack of assurance about exponents is revealed repeatedly in the variety of language he uses to read exponential expressions. *Ambiguous and conflicting uses of the language* seem to provide evidence that the underlying conceptions lack stability and adequate differentiation. One time when Dan is asked to draw a picture of four to the third, he says, "four to the third; that's four three times. Oh no, no, no four and that times four. That's four; four times four; so that's four and four and then this whole thing three times." It appears that his undifferentiated ways of describing multiplication and exponentiation such as "four three times" may contribute to his difficulties in distinguishing the required operations.

Exponential Expressions as Representations of Numbers. In the course of the interviews, Dan expresses a second insight which can be described as *exponents and exponential expressions do NOT behave as "numbers."* As Dan said:

> 9^4 isn't a number, like a regular [one]; 9^4 is just a way of representing nine times nine times nine times nine. 9^4 is really representing 729 times nine, whatever that is, and 9^1 is way of representing nine--, so really, this isn't numbers that we're working with here. It's just like, like symbols to represent it, I think? So it's like you can't do mathematical calculations with it as if it were real true numbers. Because it's just like kind of representing? It's just kind of like a, like a method? Like convention. They, they kind of said that this is the way that we, we, an easier way to represent; it's kind of like scientific notation--; an easier way to represent the --; so it's not like, it's not like, that would be like saying that scientific notation, 6.02×10^{23} is equal to 6.02. That's not true. I mean, that would be kind of like the same thing you're saying here. That you're treating this here, the

powers as if they were numbers. Whereas they're just affecting this number.

Slightly later he adds, "It's kind of easier with a four, like, 'cause you know how many times to multiply it out. But it's deceiving because the four (of 9^4) wasn't meant to be a part of the um, one of the numbers in the equation."

We interpret this passage to indicate that Dan is experiencing different levels of comfort with different sorts of numbers. His prototype seems to be the natural numbers written in decimal notation. Hence, 9^4 is the number which is the product, 729 x 9, and 9^1 is the number nine. He sees these natural numbers as objects, entities, not as the product of constructions themselves. Thus he differentiates this concept of number from expressions which require action on them to transform them into natural numbers: 9^4 is a "way of representing," a "method" or a "convention." He knows it can't be ignored as in the case of 6.02×10^{23} being unequal to 6.02, but still it is incomplete in that the action has not been carried out, and the number is the product of that action if it appears in integral form.

We suspect that Dan would not limit his conception of numbers to whole numbers, but would accept decimals most readily because they can be assimilated easily into his whole number conception. However, we have no evidence on his treatment of fractional formats.

We postulate that, in mathematics, not nearly enough attention has been drawn to the importance of such a "negative" insight, that is, an insight which acts as a warning signal to "proceed with caution." Although at this stage he associates an action with positive integer exponents, he cannot describe the bounds of the numeric behavior of exponents and exponential expressions; he recognizes only that they do not always behave as expected. Thus, although Dan has learned to act according to the rules for manipulating exponents when given in recognizable rule-governed form, the instances which are novel to him are far more prevalent than one might imagine. In these, he seldom appeals to their underlying operational character. One example occurs when he reduces $10^6/ 10^8$ to $10^3/10^4$. He must learn to work within the boundaries of his understanding of exponents rather than to routinely apply his whole number methods to all instances. His attempts to construct those boundaries account for much of the first interpretative framework produced in this analysis.

In summary, contrary to the portrayal in the traditional curriculum, the transition from positive integral exponents to rational and negative exponents is not negotiated simply by the application of a set of rules for operating in the limited context of the positive integral exponent and then extending the domain to include zero, fractions, and negative exponents in a consistent manner. Dan encounters a variety of obstacles in the process; these include the resolution of the behavior of exponential expressions with prior knowledge about negative numbers and fractions and about the use of language. He seems to expect, since exponents appear symbolically as arithmetic numbers, that his familiar knowledge of those numbers should guide him in his actions. In these instances, he makes little appeal to their function as counters of the operation of multiplication, or to any meaning he may derive from the extension of the counting concept to negative and fractional exponents.

We find that neither a description of Dan's activities as rule-governed nor as governed by competing frameworks adequately accounts for his actions and statements. We find that he holds certain expectations for the value of a number related to the form of its representation (positive, negative, fractional) but has difficulty coordinating this with the form of an exponent and with the value of an exponential expression. Exponents symbolically appear to be numbers, but their form is not consistent with the value of the exponential expressions. Much of the effort in the first framework goes into establishing the boundaries on the "integral" behavior of the exponents, and this process seems best described by a developmental model based on problematics, conjectures, and refutations or resolutions which are event-oriented and selective.

Framework Two: Exponential Expressions and Local Operational Meaning

The second interpretative framework involves Dan's attempts to reestablish an operational character for exponents including the non-positive values. In doing so, we witness two types of attempts. The first involves creating an operational meaning for negative exponents by viewing them as the "opposite" of a positive exponent, *operationally* opposite, as opposed to his previous conjecture of being opposite in magnitude. We have labelled this *local operational meaning.* The third interpretative framework will describe how he begins to generalize on this insight and coordinate it with his other schemes of operating to lead towards a more global, systemic, and operational meaning.

His second attempt involves his engagement with the construction of what mathematicians recognize as an isomorphism between the multiplicative structure of the exponential expressions and the additive structure of the exponents. As is most easily recognized with logarithms, exponents act to relate two corresponding systems of numbers and *the exponent reduces the operational complexity*. Thus, if the bases are the same, the operation on the exponents is addition when the exponential expressions are multiplied, and the operation on the exponents is multiplication when a number (or exponential expression) is raised to a power. As is well known, it was this characteristic simplification in calculation which led to the development of logarithms (Hogben, 1937; Flegg, 1983).

Local Operational Meaning. In the first framework, Dan comes to understand that exponential expressions cannot be mindlessly treated as whole numbers due to their incompleteness. The beliefs seem to indicate a vague awareness that exponents do not measure objects but that they keep track of the repeated multiplication. He seems to be aware that they cannot be added when the operation on the expression is addition. In sum, exponential expressions act in unexpected ways. However in this framework, we analyze one nearly successful attempt Dan makes to establish a meaning for negative exponents based on his initial definition. This analysis provides us insight into the complexity of coordination of schemes that is required to extend the domain.

Reversibility (Krutetskii, 1976) plays a significant part in explaining primitive extensions in operations, as will be evident in the following two excerpts. Dan is asked if he can create a meaning for 9^{-2}. He responds: "I don't know. Like nine; nine times nine; 9^2, and then 9^{-2}; kind of reducing it, breaking it down. But I knew division wasn't right, just didn't seem to fit in."

Later in the same interview, the interviewer asks if he can construct a problem whose answer is 10^{-2}. The intent of the interviewer was to see if he would respond by giving a problem such as 10^3 / 10^5, since he knew the restricted case rule for subtracting exponents when dividing exponential expressions of the form $a^x / a^y = a^{x-y}$ when $x > y$. He could then work such a problem by multiplying out or cancelling names for one. He responds by attempting to extend the "abbreviation" meaning of the exponent into an operational meaning. His spoken words express his sincere attempt to extend that meaning in light of his experiences that day. He says:

> So it's just a way of representing it then. It would be ten less, by the number ten which is being made less of itself by a factor of -2. So somehow relating them along with another ten and having it, instead of like being increased, where if it was positive, it would be times ten times ten because it's increased by a factor of two, but you want to decrease it by a factor of two. So to decrease ten by a factor of two using just these two numbers, ten and ten. Hmm, 10^{-2}. So that's where I stopped. I don't know.

He writes 10/10, since he has rephrased the problem as decreasing ten by a factor of two using only these two numbers. If he had not assumed that he needed to start with 10, as opposed to starting with one, he might have been successful in this attempt. However, for him to do this, he would have to recognize that the number *one* is the origin from which 10^2 is created by repeated multiplication by 10 (1--> 10--> 100), and hence 10^{-2} is formed as 1--> 1/10--> 1/100 by repeated division by 10.

This seems to be a particularly difficult realization since the parallel case in repeated addition starts from zero as the appropriate origin. Conceptualizing the multiplicative identity, a unit, as an origin for multiplication requires a deep measure of insight and a revision of the frequently-held tacit assumption that you always start at zero (the origin). If he knew that raising a number to the zero power gives one, this would be a possible entry into this idea. However, since knowing that may be argued to be dependent on finding the appropriate origin, there is circularity in that suggestion.

In sum, Dan seems to show evidence of localized attempts at creating a meaning for negative exponents. He seems to readily accept the legitimacy of such localized reversals of operations. We believe that this action on his part represents significant progress, for he has attempted to move beyond his bounded definition of positive exponents in an operationally grounded way and attempted, in doing so, to coordinate the symbol manipulation and the operations which drive them. The significance of this framework lies in its potential to highlight three observations.

1. It is essential for students to engage much more seriously in the *operational qualities of exponential expressions.* The traditional rule oriented approaches seem to emphasize how to move expressions around without enough focus on the operations which underlie that movement.
2. Even this simple attempt highlights the coordination of schemes which are necessary to gain insight into an isomorphic relationship. Dan must simultaneously recognize that a repeated division is the inverse operation to multiplication and that the use of the negative exponent signals division of magnitude two (i.e., two repeats).
3. Such insights are particularly unlikely for Dan to achieve because the system of multiplicative relationships (1/100, 1/10, 1, 10, 100....) is not well differentiated from the more commonly experienced counting world or the world of integers. Thus when, in the first interpretative framework, Dan expresses his belief that exponential expressions are not numbers, he is expressing a belief that all numbers must act like the counting numbers.

Framework Three: Exponents as Systematically Operational

Hawkins (1974b) defines a concept called "extending the domain" using a quote from C.S. Peirce, "a mathematician often finds what a chess-player might call a gambit to his advantage; exchanging a smaller problem that involves exceptions for a larger one which is free of them" (p. 116). To create a meaning for negative and fractional exponents, a student must extend the domain of allowable exponents. It is typically assumed that this extension occurs by a simple adoption of his/her previous definition of exponents as abbreviations for repeated addition. It is also assumed that, since the student's previous definition cannot be applied directly to the case of negative or fractional exponents, the student will easily accept the formal definition ($x^{-a} = 1/x^a$ and $x^{a/b} = \sqrt[b]{x^a}$). However, our data challenge this assumption and suggest that such extensions pose critical challenges to the students. Even when these extensions are grounded in the rules for exponents, students experience difficulty in accepting the broader generalizations.

As we examined the tapes, large segments of conversation were involved in Dan's attempts to create such an extension of the domain which allowed him to grasp the systematic relationships among the multiplicative/ divisional relations on the exponential expressions and the additive relations on the exponents. Describing some of the components involved in such a construction is a difficult challenge, and at this writing, our understanding of this process is tentative. We can offer some pointers towards interesting episodes and ways of framing his statements and actions.

As we entered the interviews, we expected students to be relatively easily convinced by the systematic consistency of two methods. The first was to assist the student in creating a chart with the exponential expressions (9^4, 9^3, 9^2, 9^1, 9^0, 9^{-1}. . .) in the left column, the repeated multiplication problems

(9x9x9..) in the middle column, and the products (729, 81, 9. . .) in the right column. The expectation was that the students would see the pattern in the right column as division by 9 and extend it to predict the appropriate values for zero and negative exponents. The second method involved using the basic rules of positive exponents and applying them to create plausible meanings for negative, zero, and fractional exponents. For example, $3^4/3^4$ can be argued to be equal to both 3^{4-4} and to 3x3x3x3/3x3x3x3. Thus, $3^0 = 1$. Negative and fractional exponents were introduced through similar arguments.

Initially, these forms of arguments to extend the domain were simply not convincing to Dan. Perhaps some of this difficulty stems from the fact that his understanding of the basic operations appears to be relatively unexamined. Thus, in working with the extension of the table, he can predict movement up the chart easily, suggesting to multiply 729 by 9 to get 9^4. Next he choses to work with the middle column, listing the nines. On 9^1, he spends some time discussing what to write in the middle column after the times sign. He tries "nothing" and then writes in a 1 as a place holder and settles on 9x1. He claims that he can put nothing in the middle column of 9^0. The interviewer prompts him to continue the pattern in the third column. He stumbles again and has great difficulty identifying how to get from 729 to 81 (and suggests division by 81), sees with slight hesitation that he gets from 81 to 9 by dividing by 9. He says "nine into 81 goes nine. And nine divided by nine is just one. Hmm. But I don't like to think of it this way."

It seems that a chart such as this should be quite powerful in teaching exponents, but in this case, Dan is unimpressed. One question we are left with is: "are there ways to work with and present such a table that would be more successful?" The interviewer decides to shift and to approach the problem through a division problem and asks him what $9^5/9^3$ equals. He replies easily, "It would equal nine to the--, 9^2." His reason is given as "five minus three." He adds spontaneously, "Ah, now this, this really ties back into here somehow, now. I know because division and, alright, division and subtraction. Why don't I break this up?" He writes out $9^5/9^3$. When asked for a problem whose answer is 9^0, he responds, "Then, I, I--, my first impression would be $9^5/9^5$, but that's equal to one, which is equal to 9^0, which is what you've been asking, right?"

Though he finds this "definitely" more convincing, Dan expresses doubt about extending the domain through these global consistency arguments. He says, "I'm more convinced because everything works out neatly and orderly here, but I don't, I'm not convinced in that, um, it's kind of like being convinced in something because, say like, I would be convinced in some complex scientific theory, not because I understand the theory of relativ--, relativity. I wouldn't, uh, understand it any bet--, I mean I wouldn't have any faith in it any better because I understand it; obviously I don't understand how it works, but, because someone told me, Einstein or someone, someone told me that such and such works out, then I 'd be able to kind of like, I don't know how that related into here. But I mean, I don't see, I can't understand, well, I kind of, a little bit, but I can't understand, I can't correlate these two ($9^5/9^5$ and 9^0).

Later he offers a similar reflection: "Well, I see that it's all kind of a big cycle, it's all connected and you can put it in order, different orders and it's all logical. It all works out, but I can see that if I put some time into it, I would be able to figure it out since all, it is all kind of logical. But right now, all it is, is while I'm looking at it, it, it makes sense, you know, the order of them all, they all make sense but they don't make sense, when I, you know, you know. If you didn't tell me this, I wouldn't be able to write it out."

These quotes indicate that Dan does not easily accept these forms of plausibility arguments. We watch Dan become accustomed to the form of argument while he overcomes his resistance to the lack of intuitive meaning for zero and negative powers. At first he accedes only to the results, but doesn't find them convincing. He attributes the insights to the interviewer and resists any claims to ownership. Not wanting to dismiss the impact entirely of these forms of reasoning, we have analyzed the course of Dan's acceptance. Over the course of the interview, increasingly we see Dan apply these "codes" without comment. When questioned, he can produce an equivalent problem using positive exponents, and he later even uses this method to create and name a meaning for negative exponents, which is evidence of giving some validity to the method.

Through this pattern of interaction, which seems dependent on the assumptions he holds about authority and received ideas, he demonstrates an increasing capacity to *accept* the conventions and to "accept" the meanings for non-positive exponents. To ignore the impact of this kind of learning--acceptance of expertise through learned patterns of behavior--would be to ignore a very significant theme in the interchanges.

Furthermore, there is considerable variation in the character of the exchanges. Near the end of the second day, he seems to make a breakthrough using the example of 10^{-2}. He tentatively supplies the example $10^6/10^8$ and, when asked if that is meaningful to him, replied, "Oh, you know that it's a big--, because you have a really small number up here and a really large one down here, you know that it's a fraction." Initially he tries to simplify it by reducing the ratio of 6 to 8 to get $10^3/10^4$. With encouragement, he writes it out in extended form and gets an answer of 1/100. He draws a connection: "So that's just like scientific notation, like one, two, into the left direction." Perhaps more importantly he draws one other connection: "And 10^2 is equal to 100 and $10^{-2} = 1/100$," and falters: "So now that, that kind of ties in with what I learned, that $10^{-2} = 10^{1/2}$." He tests that claim and, seeing that $10^{1/2}$ is slightly greater than 3, [since 1/2 is a fraction], he rejects it for a moment. (It resurfaces later.) The meaning of 10^{-4}, 10^{-3}, and 10^{-1} are explored as answers to division problems. Dan then concludes, "So it's just, just by making it the inverse that you get rid of the negative sign." He then returns to the problem of $10^{-2} = 1/100$ and $10^2 = 100$ and changes the two to a positive two and crosses out the 1 over the fraction bar; then he changes the two in the second problem to a negative two and adds a 1 over the fraction bar commenting, "Which is the same as putting the inverse."

This exchange is characteristic and significant in Dan's case. First, when Dan names a rule and labels it, the label *signifies* it for him, an action which is repeated periodically in his interviews. Later he will refer to it using similar and slightly modified language on "making an inverse to get rid of a negative sign." Also notice that, in spite of the relative naivety of knowledge of exponential expressions, he uses appropriate and sophisticated language here in *his* choice of the term, inverse. Later, in solving another problem, he recalls this episode about the transformation of 10^2 and 10^{-2}, referring to its location on the page. It might prove interesting to note also that the problem involves more than shifting expressions; it involves the movement from 2 to -2 in the exponent and the concomitant movement from a number to its multiplicative inverse. As such, it provides a rather impressive window into the isomorphism---more than one might expect from a rule such as $c^{-m} = 1/c^m$.

In our analysis, we refer to such an episode as an exemplar and discuss its impact on the interview process. *Exemplars* are moments and exchanges which seem to be unusually significant to the participants. They are marked by later references to the example, to its "name" or its location on the page. For the interviewer, they seem to reveal an important insight into the student's ways of thinking and acting. For the student, they appear as strong constructions, helping them to resolve an issue or connect ideas together.

This exemplar has another interesting feature, the spontaneous use of analogy: "This is like... ." Though for most experts, the reference to scientific notation may be viewed as a straight-forward application of exponents, we believe it functions for Dan at an *analogical* level. Place value is an application (and perhaps historically, a source) of exponentiation. Thus to students, in scientific notation, adjusting the place value in concert with the exponents is more a rule of coordinated action. At best, it is described as multiplying by ten in a local place-by-place fashion. For Dan, in this instance his analogy is likely to have resulted from the fact that the problem used powers of ten. It seems quite likely that new insights into scientific notation accompanied his insight into negative exponents. Over the course of the interview, these spontaneous analogies seem to be indicators of moments of significance for Dan.

He now seems open to reexamining the chart. He fills it in for the negative values using his new rule for inverses. He works up and down the right column and is then able to offer his own generalization. He states: "So when, when you go from higher numbers to lower numbers that are factors of nine, you divide by nine, and when you're going from lower numbers to higher numbers and those numbers are a factor of nine, then you just times by nine, and, umm, nine to a negative exponent is just a way of representing (pause), hmm, it's a way of representing 1/9, just a way of representing a fraction of that number."

In sum, for students to acquire an understanding of the systematic operational relations in exponents, they must effectively "extend the domain" to zero, negative, and fractional exponents. For Dan, this extension is attempted operationally in two ways. First he tries local extensions, which are only modestly successful. Then, with the encouragement of the interviewer, he considers more global, pattern-driven distinctions. Although he successfully creates examples and sees patterns, he resists accepting the consistency arguments as intuitive and convincing forms of argument, giving over the responsibility for them to experts or to the interviewer. However, over the course of the interviews, he increasingly appeals to them spontaneously. In one instance at least, he acts with initiative to establish a meaning for negative exponents and then generalizes on that meaning by creating a phrase to describe it. He draws an analogy to scientific notation, another action which appears characteristic in his moments of insight.

The question which drives this framework is "What does it mean to develop insight into the system of relationships between exponents and exponential expressions?" A mathematical framework for answering this question can be based on three familiar tools: rules, isomorphisms, and functions. That is, the relationship can be described as a set of rules for translating back and forth, as a more holistic isomorphic relationship which stresses the operational character, or a function showing the mapping of the one-to-one correspondences in the isomorphism.

It appears that each too has advantages and disadvantages. It also appears that few precalculus mathematical texts attempt to use the isomorphic structure, preferring to rely on the other two. The rules appear to have the disadvantage that students gain a facility with symbol manipulation and placement without examining the underlying operational qualities. Furthermore, they tend to be learned independently and are not well-integrated. The isomorphism stresses the operational character of the individual structures and the elegance of the parallel structures, but it is a difficult concept in itself to represent and communicate. Functional notation allows us to express the rules of correspondence most succinctly, but few students are taught to examine the underlying operational characteristics. We believe that an examination of the potential use of the isomorphism in creating a mapping which bridges students' manipulative skills with exponents and their understanding of exponential functions is a worthwhile undertaking.

One other topic of interest in the framework of the systematically operational meaning of exponents concerns the role of context. In order to assess the stability of Dan's acceptance of the equivalence of 9^0 and 1, the interviewer poses a contextualized measurement problem. It involves measuring a planet with a measuring stick. The interviewer varied the size of the radius of the planet and the length of the stick. The last example involves a tiny planet of radius 10^2 feet, a measurement stick of 10^2 feet, and a question, "How many sticks would you need?" Dan replies, "Then that would just be one stick. That, that's the diameter of the planet. The radius. "

Interviewer:	How would you have gotten there, explaining the calculations?
Dan:	$10^2/10^2$.
Interviewer:	And you say that you're sure that would be one?
Dan:	That would be one.
Interviewer:	Okay, what would that be in exponential notation?
Dan:	10^0?
Interviewer:	In that case, do you feel comfortable that 10^0 is equal to one?
Dan:	No, not any, it doesn't help me at all.

Note that his answer is quite appropriate. Within this context, there is no need for the notation of 10^0, since the required action is given in the answer of one stick. On another contextual problem involving negative exponents, Dan simply ignores the signs and works within that system. To design legitimate contexts which promote the need for resolving the fractional and negative exponential meanings is a challenge to be tackled in the second phase of this research.

Framework Four: Exponents as Counters

Exponents express repeated multiplication. Thus if we are to understand the multiplicative aspects of exponential functions, we must understand the concept of repeated multiplication. An exchange between Dan and the interviewer on the difference between 9+9+9 and 9x9x9 prompted us to investigate his conceptions of repeated multiplication. He had described the difference by stressing that in 9+9+9, the answer requires only one step, whereas 9x9x9 requires an interim solution. After working with these two in trying to solve the problem of how to get from 729 to 81, he comments:

> Dan: I mean, what bothered me really--, unless--, like it wasn't just like nine times nine times was equal to something. It's nine times nine--81, and then that times nine. So its kind of like broken up. So--
>
> Interviewer: So you lost part of it?
>
> Dan: You lost part of it. So all of a sudden, you're dealing with 81 instead of nines. Whereas in addition, it's nine plus nine plus nine. It's just, they stay as nines."

One interpretation of this involves both the size of the numbers and automaticity of simple multiplication and addition facts. We checked this possibility by posing a repeated addition problem with larger values and a repeated multiplication problem with smaller values. The tendency to treat them as conceptually different seemed to persist.

Thus, to look for a deeper understanding of his claim, we posed to ourselves, and later to Dan, the task of drawing a picture of repeated multiplication. At first he was asked to draw 5^3 and, when the exchange became too difficult to follow, he was asked to draw 2x3x4. Before continuing, you might wish to take a moment to sketch a drawing of it yourself.

Dan draws two times three:

Figure 2: Dan's drawing of 2 x 3.

and says, "Two groups of three." Then he revises that description as he draws and says, "Two increased by three, which is the same thing. And then this whole thing increased by four. So then this, see, I don't know how to represent this whole quantity except--, without putting parentheses around, uh, that."

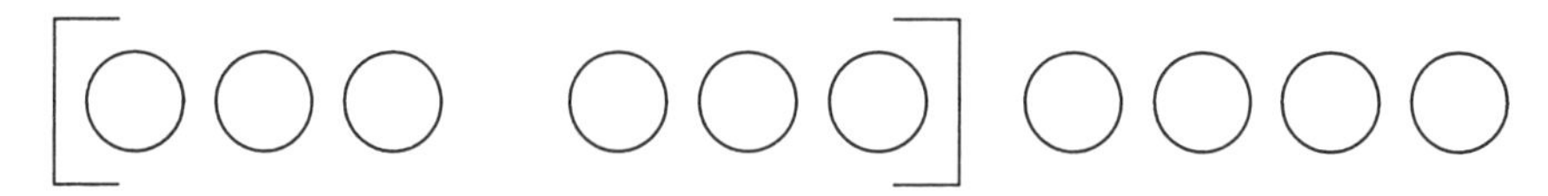

Figure 3: Dan's drawing of 2 x 3 x 4.

When he sees that his drawing gives a total of ten, he decides to start with the answer and adjust the representation to get it. He asks the interviewer's approval for "figuring out what this quantity is in pictures just by knowing that the answer is 24 and then just converting into pictures."

In this first exchange, there is some evidence of what appears to be two competing frameworks for multiplication. In the first framework, the first number determines the number of groups and the second number determines the size of each group. We name this approach to multiplication: *m groups of size n.* It can be classified with the statement: to multiply m and n is to create m groups of size n.

When this meaning of multiplication is used for a product with three factors, an interesting conflict can occur. The convention for interpretating written strings of repeated multiplication is to start from the left. Thus, m x n x p= (m x n) x p. However, if the student coordinates this order of action on the string (m x n x p) with the interpretation that multiplication is m groups of size n, a conflict may be created. This is what we witness in Dan. At the end of the first action, he has created a picture of two groups of three dots. Now in order to continue to interpret 6x4 as six groups of four dots, he needs a group of size four, not a group of size six. In order to use this concept of multiplication successfully with a written expression (m x n x p) and without redrawing, one would have to start on the right with (n x p) and work to the left.

In Dan's second statement, we see the first evidence of an alternative perspective. Dan restates two times three, as two increased by three and in this case, we take him to mean that multiplication increases the number of objects and that it should increase by four "somethings." In this case, the original group size is set by the m, and it is increased n times. Within this framework, labelled *m things increased n times,* we witness two kinds of action. Early in these interviews, Dan increases m things n times by *increasing each of the m things, n times, one-by-one.* Later, he *treats m as a group and increases it n times.* These two forms of multiplication can be applied to the string m x n x p from the left to the right. The first interpretation will also work from right to left, though it becomes tedious and time-consuming.

Thus we propose that two classes of interpretation of multiplication of m x n are in evidence in Dan's transcript. We witnessed m groups of n size and m things increased n times. Within the second interpretation, we see the m things either individually increased n times or treated as a unit of size m and increased as a unit n times. Thus, three versions of multiplication were in evidence. We further wish to emphasize that what makes this exchange interesting is that Dan must *coordinate* three modes of communication: spoken language, dot pictures, and written symbols. Using all three modes, he strives to produce the same "answer" of 24. Each of the forms has its own conventions, its own characteristic paths of enactment, and its own obstacles. It is from his records of the three that we strive to create a model of his understanding of repeated multiplication.

In the next section, we witness how the model of *m things increased n times* can lead to difficulties in deciding whether to count the original set in the final solution. For Dan, it leads to a consistent tendency to produce m too many dots.

Having gained the interviewer's permission to work backwards from the picture of 24 dots, he draws them and comments:

> So that's 24. So I have to group this up into two times three, which is six, so I'll take this and then that times four. So then this whole thing here, two times three, is equal to what I've circled here. And then this whole quantity times four is equal to everything that I left out, what I didn't circle.

If this were not an interviewing situation, he might have been satisfied with the previous explanation. However, when the interviewer asks his reasoning, he expresses his assurance with 2x3, but on (2x3)x 4 he says, "Now this--, this whole quantity times four; um, see, that's why I have trouble drawing this up here (5x5x5)."

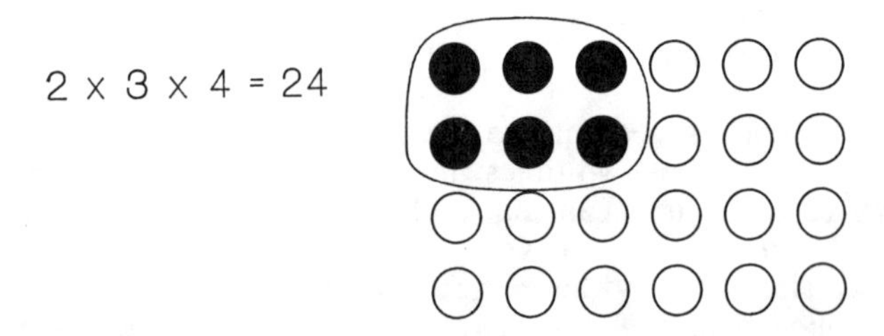

Figure 4: Working backward from 24 dots.

He colors in the six and says, "For each one of these circles that are in, I would add four to them." We take this as evidence that Dan is not treating the partial product of six as a unit, a process we label "unitizing." Instead, he wants to repeat each member of the collective of six individually and to copy it four times--what we named above as *increasing each of the m things, n times, one-by-one.*

He redraws his picture with six dots across the page and adds four more dots below each circle and says, "I would have six circles here and I'm going to take these, these circles here, which is two times three and I'm going to put them over here. And then for each one of them I'll add four to each one of them. And they will come out right, right? Will it?"

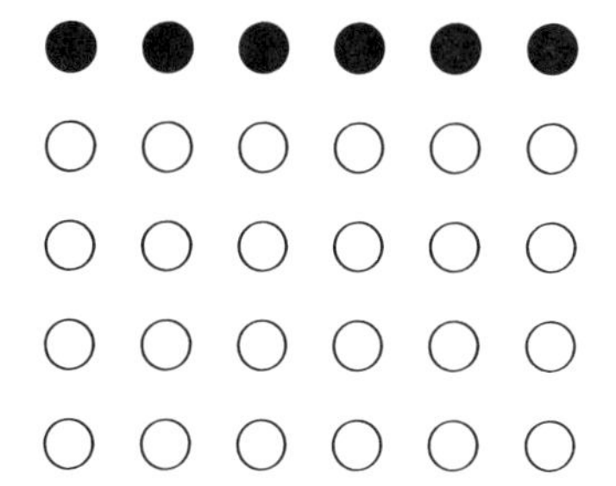

Figure 5: Increasing each of 6 things 4 times.

He gets 30 and moves back to the previous picture. The interviewer, in a rather directive way, prompts him to review how he drew 2 times 3 as "two groups of three" and, pointing to the x4 symbols, asks "Now what does this say to do?" He responds by saying, "Let's see, it takes, take this here [the picture of six] and make four groups of these." He then transfers the picture of 2x3 and draws four groups as shown below:

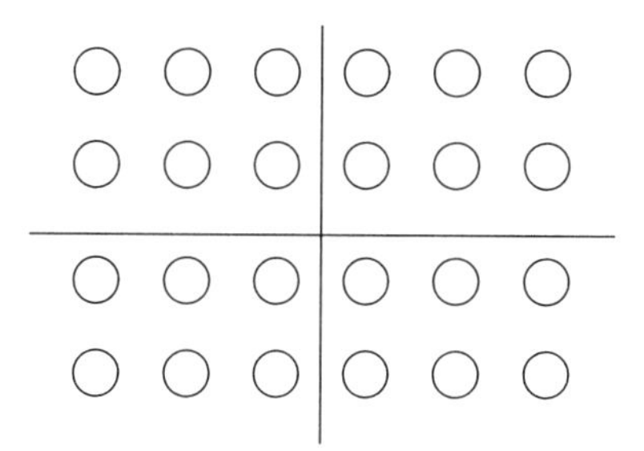

Figure 6: Four groups of 2 x 3.

This transfer keeps him from his earlier mistake of creating four more boxes (and counting the original ones). The directness of the prompt also may have prematurely encouraged him to treat the six as a unit (or as a group) and thereby influenced the change in his action to "unitizing" rather than by increasing each dot by four more. It seems plausible that he has competing constructions at this time and that furthermore, he feels little ownership of this new construction, for when he returns to his original picture, he says, "Um, wait now, how did I do this here? Anyway I--I just forgot what I just did here--, just a complete blank on this."

The issue is not resolved for him as he returns to the original drawing where he started with the twenty-four dots and tries to figure out how to segment out the other 18 dots. He says, "Because I know that I want to take out--, I want to divide this up into groups and attach it to something in here." He circles groups of four and attaches them to each dot in his original group of six but quickly sees there will not be enough to complete the process.

He looks again at his newest picture and says, "So, what I did was I took, uh, alright, I had two groups of three, and I broke that up into four groups of, two groups of three. And what I want to do is relate this back to this [returning to the first drawing]."

He tries to attach groups to each dot again and then comments, "Um, somehow I have to divide them up and I really feel like, like I'm not doing so well here because it's pretty easy and I can't think it out. Oh, alright, alright, alright. So we're taking this and we're increasing it by four times, so we're taking this--." He now circles the other three groups of six, though his language, "increasing it by four times," still contains the ambiguity about what becomes of the initial unit of six.

When he returns to the vertical columns picture (see Figure 5), he sees the difference and says, "And I took these two groups of three and I wanted to add four to it. Alright. But now, why was that wrong? I should have just added three, because, um, because I already had, because this is four already."

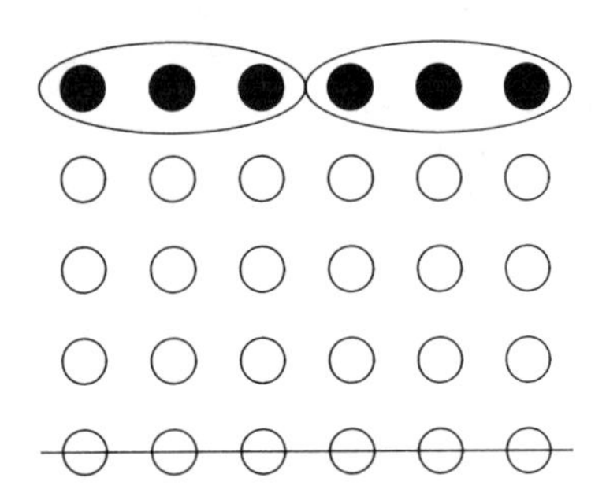

Figure 7: A correction of increasing each 6 things 4 times.

Still the issue is not resolved, for he comments, "I don't see the logic behind that, but I know that's the relationship." The interviewer prompts him to look at the last picture and asks, "Over here what you did was to do what with it [the group of six]?" He replies, "I increased it by four but I counted the one I started with. So this one, I'll count the one I started with, and then increase it by one, two, three, four."

In summary, we see an evolution of a way of coordinating the three forms of communication to create a consistent interpretation of m x n x p:

1. In written form, it is interpreted as (m x n) x p.
2. In verbal form, it is interpreted as (m x n) x p, where he (a) treats m x n as a unit, (b) increases that unit p-1 times, and (c) counts the starting group in the total.

3. In the verbal description, he uses the phrase, "increases it by p, but counted the one I started with," and yet he reproduces only p-1 times.

When he tries to draw 5x5x5, we witness how he comes to rely on this interweaving of meanings:

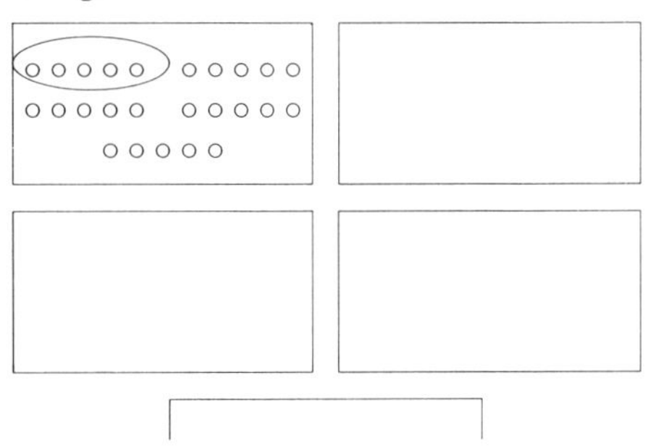

Figure 8: A drawing of 5 x 5 x 5.

He comments with relative ease:

> So that's one set of five, and that's one, and I count that one set of five, and then I've got to add four more sets of five. Now I really see what you're talking about [note that he turns over authority here] by drawing this out, because--, alright, I want to increase five sets of five so I start with this set of five and then increase that one, but I count this one.

This exchange illustrates some important features of multiplication which suggest that the role of the unit as an initial starting point for multiplication needs close scrutiny.

1. Dan is fundamentally correct in his claim that 3+3+3 is different from 3x3x3; that is, that repeated addition is not a parallel structure to repeated multiplication. Repeated addition begins with zero and each new group of three is tacked on to the previous group. There is no need to change the size of the unit. In repeated multiplication, there is either a switch in unit size or a dependence on returning to the original unit by decreasing the dimensionality one level.
2. Dan focuses on the *change in quantity* when he interprets the operation, and this later becomes the basis for examining functions, wherein he focuses on how they change. Thus he describes multiplication by m as an increase by n and sees this action of copying n times as the action which can be attached to multiplication. This leads him to overestimate consistently in his total by m. His resolution of this overestimate can be viewed as somewhat of a patch, as he learns a lesson, which he codes as a reminder, to count the first group as one of the m increases.

3. Dan seems to rely on his model of multiplication as a copy of each unit within his group whenever he encounters a new situation. Later in the interview, there is some evidence that he begins to curtail this into a process where the first factor of the multiplication is treated as a unit, and it is then copied n-1 times.
4. To interpret Dan's actions, the analyst can draw from three modes of communication used by Dan: verbal, symbolic, and iconic presentations. The model which emerges will attempt to explain activity in all three domains.

In the next interview, Dan is asked to draw 3^4.

Dan: Okay. Three times three, that's uh, three groups of three. Okay. Now then I'll write this out three more times. So I take this and now all I'm doing is, I'm able to figure out this because what struck me there was I already had one group, I just added two more. So I started out with one, but at this point, I still wouldn't

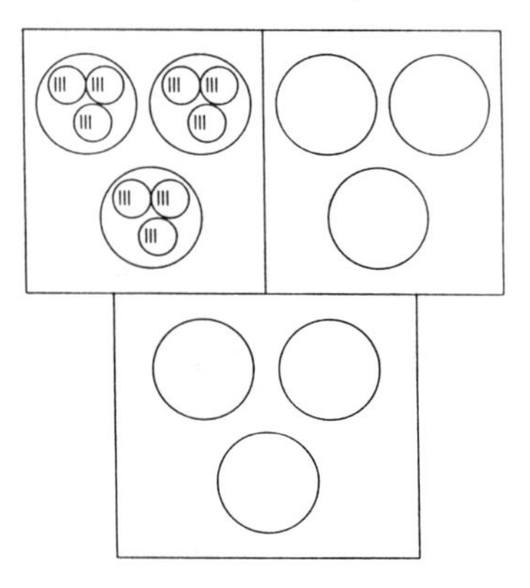

Figure 9: Dan's drawing of 3^4.

go back. So, and then I'd write this whole thing out three times, which is easier.

Interviewer: So does that come out the right value?

Dan: I'm not sure. 1, 2, 3, 4, 5, 6, 7, 8, 9 times nine; nine, nine; nine that's 27. 27 times three; that's, that's, um--, that's 81. 3x3x3x3 is 81. That equals--.

Interviewer: Okay. So then, the question is, where did the four go? Where's the, where's the four, or the three in this picture [previous problem].

Dan: I think its easier in this one. Because [??][3] Uh, four, it's not in the picture. I don't think. So I think, this just--, alright, the four here, I think that this tells me how many times to write this out. So, uh, that should be here somewhere. It's telling me to write it out four times. Uh, hmm. Unless it's lost. I don't know. Uh, let me see here, 1, 2, 3, [Draws circles around groups of nine in the other two large boxes.] See. I know it's there somewhere. I know it.

Interviewer: Okay.

Dan: But, um, oh, unless this is just, just this. No, then, this does work out. Um, I wish I knew where it was. Let's see, I'm taking three and I'm writing it out three times.

Interviewer: Here's three, right?

Dan: Right.

Interviewer: So where's three times three?

Dan: Alright. Here's three. Three times three is this here.

Interviewer: Okay so here's three. One circle, right.

Dan: Oh, alright, so you have 1, 2, 3, 4, so that, um--. [He counts a little circle and two large circles ??]

Dan: One circle, that's three.

Dan: Yeh.

Interviewer: And where's three times three?

Dan: That's this.

Interviewer: Okay, that's the bigger circle. And where is three times three times three?

Dan: That's three times three times three?

Interviewer: Uh, huh.

Dan: Is, is this here.

Interviewer: Okay.

Dan: And 3^4 is so, then [??] is just the amount of circles. There's 3^1 [points to small circle], 3^2 [points to next circle], 3^3 [points to box], and this whole thing is 3^4. Right.

This excerpt suggests the idea of the exponent as a counter. By this we refer to the sense that Dan expresses when he says that the four is not represented in the picture as a collection; it remain unseen in that form and is apparent only as the number of repetitions of the unitizing process. Thus we have a number which functions in an "unseen" fashion; it has no sense of the cardinality typically associated with numbers. In this fashion, it reinforces Dan's earliest claims that exponents are not numbers in a traditional sense. As one considers the question of what kind of number an exponent functions as, one might consider the role of the successor (Newman, 1956) in Peano's axioms. Exponents may yield to more of a successor explanation than a cardinal one.

There is a strong analogy here between the role of the exponent as a counter, and the need to see the different roles of the factors in repeated multiplication. All three meanings for multiplication mentioned earlier (none of which include the meaning embedded in length x width = area), have a parallel role for the multiplier. In fact, it is likely that it is exactly this parallelism which creates the isomorphism on which logarithms are based.

In the interviews, we explore binary searches and chain letter examples with Dan which require him to make use of the concept of the exponent as counter. As a result of these interviews, we propose that this is a very powerful intermediate notion in a student's progress towards functions. Thus we argue that if students are to gain a more powerful understanding of exponential functions, we need to stress the operational basis of those functions and to explore the intermediate value of the exponent as a discrete counter. We suspect that if the idea of the multiplier as a counter were used more in the introduction of linear functions, students' understanding of linear functions might improve, with possible effects on the introduction of exponentials.

Counters in Context. Another part of Dan's exploration of exponential expressions involved having him work on the location of planets, where their distances from each other and the sun are given in scientific notation. In this arena, he has to coordinate his diagrams and his manipulation of the numbers. The problems he works on are: "Venus is 1.1×10^8 kilometers from the sun. Mercury is half that distance. (1) How far is Mercury from the sun? Saturn is 1.4×10^9 kilometers. (2) How many kilometers farther away is Saturn than Venus?" [Assume the planets are lined up.]

Upon reading these problems, Dan immediately comments, "I like these. I like word problems." In solving the second problem, he creates an equation of:

$$1.1 \times 10^8 + x = 1.4 \times 10^9$$

He changes the right side into 14×10^8 and prepares to solve for x, but does it by dividing 14 by 1.1. Then he becomes uncertain whether it is for multiplication and division that you change the exponents or for addition and subtraction. He can't figure out how to decide on this "rule I learned to make them the same."

He struggles to express in his own words why he has "made them the same," going beyond his first attempts to remember the rule. He says:

> Yeh, I'm not--, if I sat down for a while I could figure out why, but I just know that they have to be the same so that they change the numbers themselves, so that they're in the right relationship. They're in the right; it's kind of, I can't explain it, but I mean, I know like the idea behind it is--. You just have to change those, change these numbers here to be the same, so that they're both the same. But the way you get them the same is to make these two the times 10 to the whatever the same. So when these are the same, and then the decimal point moves over one, so whereas this isn't 1.4 it's 14, *its in the same realm as* 10^8.

It takes considerable time for Dan to sort out how to adjust the exponents. Later, in this same episode, he attempts to figure out why he knows from his science background that Saturn is way far out [see sketch below], and yet the distances appear to him to be equally spaced (5.5×10^7, 1.1×10^8, and 1.4×10^9), since the exponents go 7, 8, 9. This is the only case in which he writes out the values and slowly sorts out the increasing impact of the exponents. This concept of "being in the same realm" seems to serve a similar purpose of segmenting and differentiating levels, as did the counter in the decontextualized problems, hence we include it in this section.

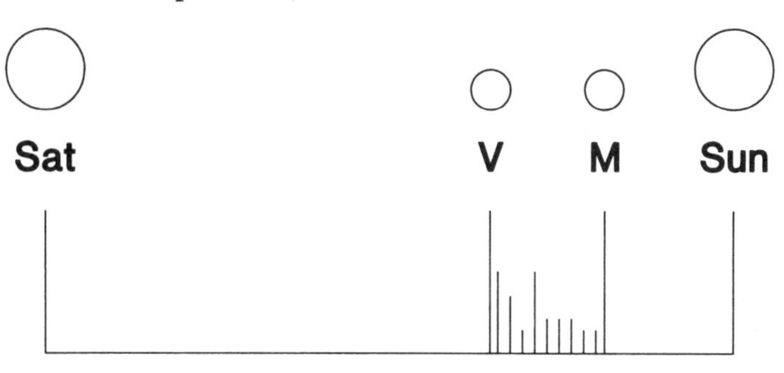

Figure 10: Dan's drawing of the planets problem.

Framework Five: Exponents as Functions

The research on students' concepts of functions relies heavily on the students' understanding of the definition of functions and the coordination of that definition with the ideas of domain, range, correspondence, specificity of rules, maximum-minimum, continuity, etc. (Vinner, 1983; Markovitz, Eylon, and Bruckheimer, 1986; Dreyfus, and Eisenberg, 1984). The question of how a student comes to understand the idea of *relatedness* between two sets of numbers or the operational character of the mapping of one set onto another is emphasized less. An exception to this comes in the work in physics by Lillian McDermott (1984) and her colleagues, where the physics content guides the formation of functional relationships, and in the work of Steve Monk (1987) where an interest in students' understanding of calculus acts as a guide.

Two additional perspectives are guiding our examination of exponential functions. First, we are increasingly intrigued by the operational basis for a particular class of functions. In this regard, we seek to examine the multiplicative structures which underlie exponential functions and how these operational relationships are embedded in different representational systems: algebra, graphical, tree diagrams, embedded squares, data tables, etc. Second, we seek to consider the role of context in relation to this operational basis and to multiple forms of representation. In another part of the project, this cross-disciplinary question is being examined through a series of interviews with scientists who work in different ways with exponentials. In the student interviews, we sought to examine these issues by providing a variety of tools, graph paper, calculator, compass, and by embedding the material in various contexts.

Very early in the interview, Dan engages with the question of why 9x9x9 differs from 9+9+9. The notion of an interim solution is one of his explanations. The other explanation is that Dan expresses a recognition, that in the multiplicative case, there is an increasing difference. He again uses this "increasing increase" on the third day to explain the difference between a problem where a banker is given $325.16 per day for three days and the work with the chart of powers of 9. In response to the question, "Do you see that as related to what we did on Friday?", he responds:

> In that (325.16, 650.32, 975.48), it's an ordered relationship in that it's increasing at some fixed, fixed rate. Whereas this rate is just increasing at, by a rate of itself. Whereas, uh, last time we met it was exponentially, you know, it increased more with, I don't know how to explain, but it increased with each increase.

For Dan, this rate of change perspective is a key in understanding how he moves towards developing an understanding of exponential relationships of growth. After the astronomy problems, Dan is given a depreciation problem: A computer originally cost $4,000. Each year it is worth 90% of the previous year's value.

1. How much is it worth after 1 year?
2. How much is it worth after 3 years?
3. Ten years?
4. Does it become worthless?
5. When is it worth less than $1000?
6. Write an equation giving its value after t years.
7. Draw the graph.

Dan has no difficulty starting problem 1. He does it in two steps by redefining the problem as "decreasing in value every year by 10%." He first gets $400, then decides he's wrong and goes back and multiplies $4000 by .90. To get the second year's value, he immediately says: "And then after two years, alright, it would be that times .90" To get year three, he repeats the process on the result of the previous value.

This recursive action on the previous value is worth commenting on. Most students are taught to work with exponential functions in the form: $f(x) = c\,a^x$. However in all of our interviews, the more recursive approach seems to be more intuitive to the students and needs to be examined first. In work with students on computers, where Basic allows the students to write something like p= p * .90, they can generate a table of values easily and intuitively by using tailwise recursion. To us, this suggests a rather different approach to functions, which may evolve a different set of symbolic representations.

When asked the value of the computer after ten years, Dan groans and comments: "I hate this type. Because, like, I'm not going to work it all the way through up to ten. I have to figure out a relationship between, what was, the value I had for two years?"

He continues:

> Yeh. Right. I'm not really sure how to do it, but I have to find a relationship of the change with each year and somehow I think it's already set up as 90% with each year. So somehow I'm going to have to combine ten years and 90% somehow into one number. And then times that by the original $4000. So I'm not sure how I'm going to combine 90%.

He then tries the conjecture that two years is 80%, etc., but finds out it doesn't agree. In all of the interviews, this question of creating a way to go from the initial value to a final value without getting all intermediate values was a stumbling block. We suspect that it may be an example of a *critical barrier to understanding* (Hawkins, Apleman, Colton and Flexner, 1982). Examining how to assist students in this transition is a critical part of the study, and so far we have witnessed two alternative paths to resolution. Both took considerable time and effort. Below, Dan's is briefly described.

Dan examines the rates of change looking spontaneously at first, second, and third differences. The pattern of exponential relationships intrigues him, but since it recreates itself, it doesn't help him. He then restates the question, "I'm trying to find what it would be after ten years," and says:

> That's the main thing I'm trying to solve. But the little thing that was trying to solve in there is how I could find some kind of relationship between two numbers. Between two years. Just between one year and try to relate that to ten years so I don't have to spend all day working it out.

He solves this problem by creating a notion of a *jump*. For Dan, a jump is how you go directly from one year to two years. He quickly assumes that this ten years would be simplified by reducing it to five two-year jumps. An excerpt from the tape explaining this idea is offered.

Dan: I would have to find the rate with one year first, before I could find two years. I would have to break it up.

Interviewer: You'd have to actually solve it?

Dan: I couldn't just jump from $4000, oh, alright, I could.

Interviewer: How?

Dan: The way I would do it is I would break it up in one year and two, but somehow I would have to change, decrease this, or increase this to adjust for this jump. I'm going to jump right from, from day zero, with zero years, we'll put that here, in which I have $4000.

Interviewer: Uh huh.

Dan: And I'll have to jump, if I wanted to jump from zero years up to two years. I would have to, since I'm skipping year one, I would have to find a way to change the rate at which it changes, because, because from with each year it's changing at a rate of--.

Interviewer: I can't read the writing, Two, $3240?

Dan: $3240.

Interviewer: Okay. So that's our object, is, could you go straight from $4000 to $3240, without having to figure out $3600 in the middle?

Dan: Yeh, I could, but...

Interviewer: How would you do it?

Dan: I'm not exactly sure how to do it. That's why I did it this way and broke it up. But if I wanted to make a jump from one year, I would have to change this, if--. Alright, this is the rate that it changes with each year. I have to find out the rate that it changes over two years.

Interviewer: Okay. What would that be?

Dan: So, if it changes--. Alright, with each year it's 90%, and then after two years, so if I start at $4000 and I went down to $3240, I would have to figure out the percentage of that change.

He follows this with a period in which he questions whether it will be 81% for any two-year jump. He thinks that the percentage rate will change at first because of the changes in his chart of differences, but then he recalls that it stays 90%, so he's convinced of his conjecture. When he tries it and gets the same value he listed for four years, he is pleased.

To this point, Dan sees no connection between the two percentages. When asked to find a rate for three years, he uses the same process, putting the value for three years over the original $4000. When the interviewer asks him if there's a way to predict the percentages without the values, he makes a chart of the size of the jump and the percentage rate of change. Again, he looks at first and second differences.

Dan does not seem to see any way to get from the initial point at $4000 to the value for ten years through his process of calculating rates of change. He gets stuck in his circular reasoning, needing an answer to calculate the rate of change. His analysis of patterns in the chart is limited to taking first differences through *subtraction*, which precludes him from discerning the constant *ratios* among the rates.

The interviewer, in an intentional effort to get him to try another approach, asks him how he would work to get the value for two years on his calculator. He says "That would be that number, 4000 times .90 and I would take that number, whatever it was, and times that by .90."

Interviewer: How would you get to the end of the third year?

Dan: I would take that number and times that by .90.

The interviewer asks him how he would write that out without writing any interim values. He writes:

4000 x .90 =

Then he starts to write:

x .90

and the interviewer says. "I don't want that number. You have to start with $4000. That's the rules of the game now."

> Dan: O.K. I start with 4000 and times by--. Oh, okay. Nine times nine is equal to 81. So I would have to take .90 times .90.

He follows this with an exchange where, as he generalizes this to ten years, he tries to multiply .9 x 10 getting 9. He does succeed quickly in sorting this out and calculating 10 years as $(.90)^{10}$ x 4000. The instability we witness in both his language and notation using the exponentials in the episode are evidence of the slow progression towards stable understanding. He summarizes his process by saying:

> Dan: No, no. So what I did is--, I just found--, I'm not sure if I could do this again. Um, but I found the change with each year, and I found the change over two years, and I found out how to get from a change, a jump of one year to a jump of two years, and I just times the percentage, .90 times .90. I squared it, but then I brought it up to the tenth power for a change of ten years, alright.

We find the transition from working recursively with exponential functions toward working with an "initial-final" view which takes a single initial point and produces a final value to be an epistemological obstacle (Brousseau, 1984b) for students. To overcome this obstacle, students need to have a strong operational understanding of multiplication. This view of repeated multiplication is not simply reducible to repeated addition. Therefore to integrate their understanding of functions with their operational insights into repeated multiplication requires a powerful act of reflective abstraction.

In Dan's case, he works with percentage rates of change and quickly hypothesizes and tests a claim that, for equal time intervals, the percentage rate of change is constant. His method is circular, however, since it requires him to calculate the value to get the percentage. He tries to use tables of values, but he works only with difference patterns which recreate the exponential relation. When the interviewer poses a form of the task which denies him the use of the recursive method, he sees how to generate his percentages. However we suspect that without having identified his own method of getting the percentages, it is likely that his recognition of the use of the exponential expression in the function would have been lessened.

Other patterns and issues evolved in the development of Dan's concept of exponential functions, but in his interviews the rate of change, the concept of a jump, and the use of the calculator were distinctive. Some of the other issues which are not discussed in any detail in this paper, but which will be presented in other work on this case include:

1. Dan's subsequent work on a contextual linear problem, which allows a comparison of his methods on linear relationships vs. exponential relationships;
2. Dan's first and last interviews, which involved sorting combinations of pictures, diagrams, and graphs and discussing three different images of functions;
3. Dan's work on an embedded squares task with a geometric sequence presentation of functions; and
4. Dan's choice of forms of representation. On these problems, he relies heavily on the calculator and only occasionally uses algebraic notation, graphs or diagrams.

Conclusions

In summary, it seems important to begin by mentioning our surprise in the course the interview took. We found ourselves much more involved with the operational character of exponents and in the multiplicative structures created by repeated multiplicative setting. In order to understand Dan's work with functions, we created much more explicit connections between what we had assumed were routine manipulations on exponents and his deeper understanding of exponents. Furthermore, we became significantly engaged in the question of how one creates an understanding of a system of relationships such as is apparent in the isomorphisms described earlier. We challenged our assumptions that students behave in a rule-governed fashion and found a need to track much more closely and carefully the kinds of exemplars on which the students ultimately rely to be convinced of these systematic reasonings.

As always with this work, we were both dismayed and awed by Dan's performance. He demonstrated that the diversity in student understanding is dramatically underestimated in most descriptions of students and that his foundations in operations were disturbingly unexamined and poorly differentiated. He also demonstrated profound insights at times, forming interesting connections and spontaneously drawing analogies. Over time, it became apparent that his route through the material was punctuated by creating language to signify important insights and by particular examples on which he took the responsibility and worked out a genuine problematic to which he returned time and again. He learned to learn more and more effectively from the interview setting by slowing down, becoming more concise in his explanations, and following up on previous attempts in order to resolve them in his mind.

Discussion: Implications of Findings for Mathematics Education

In the beginning, I classified this research as an extension of the constructivist research paradigm as exemplified by the work of Steffe and von Glasersfeld, with their own ancestry in the work of Piaget. I indicated that the extension of such work into more sophisticated mathematical content promised some intellectual intrigue and reexamination. I am now in the position to discuss some of these implications more directly.

The first observation is that carrying out investigations on students' understanding at the secondary and post-secondary level has the potential to raise eyebrows of some of the members of the mathematics community, because it challenges not only the traditional presentation of these topics but argues for a kind of psychological analysis of mathematics which might seem unnecessarily fanciful, a romantic interpretation of students' potential. Those mathematicians (and mathematics educators) who have perfunctorily dismissed the possibility of educating most of our population into an understanding of higher mathematics are unlikely to provide a sympathetic hearing to the claim that, through examining students' conceptions, we can derive legitimate mathematical content. In spite of this, with the support of this presentation of the data in Dan's case, I do make such a claim.

Furthermore, one analogy to this situation might make this claim more appealing. In *The Blind Watchmaker*, Dawkins (1987) offers a lovely description of how bats use echos to see, and in so doing, he emphasizes that their form of hearing is perhaps more appropriately likened to our form of vision in its complexity and richness of image. He describes a story by Donald Griffin about what happened when Griffin and Galambos first reported on the bats' echolocation in 1940. A distinguished scientist was thoroughly indignant and outraged by the proposal that bats could navigate by a means in which humans had just achieved a technological breakthrough. He found the proposal "implausible and emotionally repugnant." Dawkins explains such a response sympathetically and wrote:

> It is precisely because our own human senses are *not* capable of doing what bats do that we find it hard to believe. Because we can only understand it at a level of artificial instrumentation and mathematical calculation on paper, we find it hard to imagine a little animal doing it in its head. Yet the mathematical calculations that would be necessary to explain the principles of vision are just as complex and difficult, and nobody had ever had any difficulty in believing that little animals can see. The reason for this double standard is, quite simply, that we can see and we can't echolochate (p. 35).

Can the student be likened to the bat? Can we conceive of the possibility that our education teaches us to silence certain avenues of pursuit which we can find evidence of in listening to "less" sophisticated problem solvers? Is it also possible that, by learning to listen to students, we will discover in ourselves ways to map our sophisticated knowledge onto the rich and diverse passageways and dead ends (in part from lack of encouragement) of the students? And that such an experience may enrich us mathematically?

I believe that this work is significantly different from traditional research in mathematics education at this level, being more compatible with and complementary to the considerable contributions of those working on early number concepts. As opposed to younger children, students at this age are more verbal and perhaps more self-protective. We enter into an analysis of communication patterns among adults and this can deepen our understanding of education at the higher levels. In terms of the mathematics, the context also provides some newer opportunities. The material can be embedded in more variegated multiple representations, and we can examine how students move among them and make choices. We can look for consistency or lack of it across these areas.

As opposed to traditional approaches, particular examples become exemplars, and the students come to know individual problems in depth, rather than attempting to master a class of exercises with only minimal contextual meaning and with only trivial distinctions between them. Language which labels and reminds us of a significant insight seems to take precedence over the formal definitions, and definitions evolve as the student is challenged with counterexamples and extensional settings. We can begin to explore what it means to understand a system of relationships.

One final example of how our subject matter meanings are interwoven with our ways of viewing the world will be offered so that the reader may consider the potential of how our experience and our organizing frameworks interact. Imagine first what you would see if you point a videocamera at a monitor to which it is broadcasting, simultaneously.

The set of blank monitor screens that would be seen as embedded squares can be represented in two different ways by short programs, assuming each takes a simple subprocedure which is called "square: side" and which draws a square of the length "side."

```
10 To Embed Square :side
20 side = :side * .75
30 square: side
40 go to 20
```

```
10 Initial
20 Let x = 1 to 100 step 1
30 side = initial * (.75 x)
40 square: side
50 next x
60 end
```

Each of these representations has its own strengths and weaknesses. With the recursive approach on the left, we have the problem of the lack of an ending, but the recursiveness has the power to invite us to continue; it draws us in in a very pleasant fashion. The second approach, an input-output model, leaves us at a distance. Though we can quite readily see its application, we are left in more confusion to explain how it keeps on working. It does not draw us in, but requires us to predetermine the appropriate domain for x and in that way feels less exploratory.

Perhaps these two forms of describing our experience with the camera and the monitor can be useful in representing our own experience with learning and teaching mathematics. Human beings seem to be somewhat like the first approach, self-reflective and turned inward. Our understanding of our understanding seems to provide the characteristic turnings which allow us to bootstrap ourselves into a state of knowing. In contrast, we are also input-output machines in a way, who depend on others to specify the domain of our tasks and to limit and bound our entry into new worlds of subject matter. As others specify the domain, we respond with outputs which allow the others to guess at our inner workings.

Notes

1. This chapter was supported under a grant from the National Science Foundation, Grant No. MDR-8652120.

2. The author wishes to express her appreciation to Erick Smith and Julian Fleron for their thoughtful suggestions offered on the many revisions of this paper.

3. Brackets around question marks indicate untranslatable sections or comments in the interview.

9 Constructive Aspects of Reflective Abstraction in Advanced Mathematics

Ed Dubinsky

The notion of reflective abstraction was introduced by Piaget and, over a period of many years in a number of works, he expanded and elaborated this concept. He considered it to be the driving force of the (re-)constructions involved in the passage through the stages of sensori-motor actions, semiotic representations, concrete operations, and formal operations (Beth & Piaget, 1966, p. 245). But he also felt that reflective abstraction was critical for the development of more advanced concepts in mathematics. In his viewpoint, new mathematical constructions proceed by reflective abstraction (Beth & Piaget, p. 205). Indeed, it was for him the mechanism by which all logico-mathematical structures are constructed (Piaget, 1971, p. 342), and he felt that "it alone supports and animates the immense edifice of logico-mathematical construction" (Piaget, 1980a, p. 92).

In Dubinsky (1990) I tried to trace through a number of Piaget's writings on reflective abstraction and to summarize the various aspects of what has become a broad and complex collection of ideas. I also tried to concentrate on what I call the constructive aspects of reflective abstraction and to contribute to a flushing out of the role that these constructions play in mathematical knowledge and its acquisition at the more advanced levels to which Piaget referred but did not investigate very extensively. Roughly speaking, Piaget's notion of reflective abstraction has to do with (mental or physical) activities of the subject. He always considered that reflective abstraction had two parts. One is a projection of existing knowledge onto a higher plane of thought and the other is the reorganization and reconstruction of that knowledge to form new structures. As a teacher of mathematics, it seems to me that it is necessary to understand something about that reconstruction and to apply it in designing instructional methodology. It also seems necessary that, as did Piaget, we should seek this understanding through observations of students as they are trying to understand mathematical concepts together with a theoretical development to organize and interpret these observations.

I am trying to contribute to an understanding of how students construct those mathematical topics which, in the educational system prevalent in the United States, tend to appear at the end of the high school and throughout the undergraduate curricula. It is my intention in this chapter to say something about that work, and in particular, to indicate how my theoretical analysis is related to what I have observed to be happening with my students. Thus, in section 1, I present a brief description of a developing theory of mathematical knowledge and its acquisition with particular emphasis on more advanced topics than were studied by Piaget or are generally considered in contemporary research in mathematics education. In section 2, I describe the specific methods of construction that I have observed in students and give some excerpts of interviews in which I infer that these methods are being used or, alternatively, in which I attempt to explain students' difficulties in terms of their absence. My research program attempts to implement the aforementioned coordination of observation and theory. It also attempts to incorporate a third ingredient--my own understanding of the mathematical concepts being investigated. This is the subject of section 3. Finally, in section 4, I analyze three specific mathematics topics which I have studied more or less extensively from this point of view (induction, quantification, and function) and also describe briefly how a number of other topics might be understood in these terms.

A very important part of my research program is concerned with designing instruction based on the understandings that result from these investigations. In particular, I have been using computer activities to create problem situations and to induce students to make the explicit constructions that seem to be useful in acquiring various mathematical concepts. In this chapter, however, I restrict my considerations to the epistemological basis for these pedagogical developments. Discussion of the latter can be found in reports of the individual investigations (Dubinsky, in press *a*, *b*).

Mathematical Knowledge and its Acquisition

The reader should not infer from the above title that anything like a definitive statement is to be offered. My thoughts on mathematical knowledge and its acquisition are developing and will, hopefully, continue to develop. About all that I can say is that what I now think seems to me a little more reasonable than what I used to think. The interactions at this conference have had a salutory and stimulating effect on this development.

I take the view that knowledge and its acquisition are not easily distinguishable, if at all. This seems to imply that one needs to consider epistemology and learning together, and I am comfortable with that. Indeed, my present, tentative, working statement goes something like the following.

> An individual's mathematical knowledge has to do with her or his tendency to respond to perceived problem situations

by (re-)constructing (new) schemas with which to deal with those situations.

The parentheses around "new" is meant to suggest that although sometimes new schemas, very different from what have been used previously, must by constructed, it also happens that existing schemas are re-presented and used in a new problem situation and the only construction is the link between the old schemas and the new situation. Since no two problem situations are exactly the same, one can thus conclude that there is always a construction. Moreover, the construction of a schema makes use of existing schemas in ways that suggest that *reconstruction* may often be the more appropriate term.

In any case, the above capsule statement expresses my present understanding of the dynamic interdependence of mathematical knowledge and its acquisition. If you want to find out something about the mathematics that a person knows, about all you can do is try to put her or him in a problem situation, observe what happens, and attempt to infer something about the schemas that the subject seems to be using. But since the subject's responses involve (re-)constructions, you are not only seeing what knowledge is "possessed," but also observing the process of its acquisition. Moreover, it must be emphasized that mathematical knowledge is only a *tendency* to respond in a certain way. This is another reason why the inference we make will always be tentative.

I see this, not as a confusion, but as a synthesis or (to use a term in a sense that will be important below) a *coordination* of knowledge and learning. Indeed, it is not clear to me, at least as far as mathematics goes, that a person has very much in the way of "knowledge" in the sense of things that are kept somewhere and selected, like carpenter's tools, when needed. I am more comfortable with the idea that a person will have a tendency to use certain kinds of schemas in certain kinds of problem situations. Neither the schemas nor the problem situations can be specified precisely. Moreover, the fact that a person uses one schema in one situation is far from a guarantee that a similar schema will be used in what appears to an observer to be a similar problem situation, even when the two situations occur in close temporal proximity.

Such imprecision and uncertainty does not provide encouragement for what I attempt in this chapter. Nevertheless, my experience as a teacher and the reports of others suggest that this is the way people are. So, rather than try to impose an artificial structure of predictability on mathematical understanding, it is perhaps better to try to understand its unpredictability.

Given these caveats, I will now try to say a few things, in the context of mathematics, about problem situations, schemas, and responses. It might seem strange, in a mathematics context, especially with respect to mathematics education, to suggest that one needs to pay attention to the nature of problem situations. After all, in some sense, problems are what mathematics education has always been about. But the poser of a problem cannot be too sanguine about the problem that the other is responding to.

For example, a high school teacher might present her or his students with the following mathematical statement.

$$5x + 7 = 2x - 15$$

For the teacher, the problem might be to understand that 5x + 7 and 2x - 15 are two expressions and that one can imagine making various selections for numberical values of x after which one can obtain, by substitution, values for the two expressions. Moreover, the = sign is an indication that when it is done, the two values are the same. Finally, the problem includes the understanding that this equality may not always be achieved and that "solving the equation" consists of finding those selections of values for x (if any) for which the equality *is* achieved. The teacher might consider that a second part of the problem is to apply some standard procedure to actually find those values. Procedures for doing this depend on certain properties of equality with respect to arithmetic which, when applied in this situation, lead to the conclusion that -22/3 is the only such value for x. For the student, the problem might be quite different. In many cases, the problem situation as perceived by the student is to get the "right answer", -22/3 in this case. This can be fleshed out to a situation in which a number of procedures or actions must be selected and applied to the string (which is what the equation is for many students) to eventually translate the original equation into something like,

$$x = -22/3$$

Many difficulties in education can be explained by an appeal to such a difference in the perceived problem situation. For example, a teacher may well be puzzled that students who are quite good at getting these "right answers" are at a complete loss to deal with what happens when they apply their procedures to equations like

$$5x + 7 - 2x - 22 = 2x - 15 + x$$

and obtain

$$-15 = -15$$

or to

$$5x + 7 - 2x = 2x - 15 + x$$

and obtain

$$7 = -15$$

Such students may have responded in the past to the problem situations as they saw them and constructed effective schemas to deal with them. They may not, however, have responded to the problem situation as seen by the teacher and so their schemas had no facility for dealing with equations that are solved by all numbers, or by no numbers.

I might add that, in my experience, the inadequate communication between teachers and students on the nature of the problem situation is not easily (if at all) eliminated, no matter how careful is the teacher in expressing the problem or how much is given in the way of instructions and explanations.

One last point about the source of problem situations that is important, both for young students and more mature mathematicians. It is not necessarily the case that a problem situation always comes from an individual's environment or from the activities of another person. Aside from the fact that the perceived problem situation in any case derives in part from an individual's interpretation of an experience, it can also happen, especially in mathematics, that the source of the problem is largely within the individual. Often, for example, in doing mathematical research, the most important part of an investigation is to ask the right question, after which it can appear quite straightforward to re-present existing schemas and, with more or less construction, use them to respond to the question.

So what, (you should excuse the expression) exactly, is a schema? I will respond to this question very briefly here and return to it later. In the formulation I am working with at the present time, I consider a schema to be a more or less coherent collection of cognitive objects and internal mental processes for manipulating these objects. Included within a schema are a number of activities and procedures that can be used in problem solving and may become part of subsequent constructions. Although I shall speak of an individual having, or possessing various schemas (or not), I do not intend to suggest anything other than a tendency which an individual may have to respond in ways that can be described in terms of applying a schema to organize, or make sense out of, a perceived problem situation.

Which brings us to responses. Of course it is easy to say, in the context of the above discussion, that, when an individual perceives that he or she is in a problem situation, then he or she will respond to that situation by re-presenting (consciously or not) one or more schemas and using them to make sense out of the situation. Unfortunately, that does not say very much since the schemas selected can include ignoring the problem (answering "I don't know," remaining silent, leaving the paper blank, etc.), applying existing schemas either inappropriately or appropriately (in the view of the observer), or constructing new schemas and using them to deal with the situation.

This makes the spectrum broad enough to ensure that every response will be included somewhere, which is not very interesting. The serious question for mathematics education research is to understand something about what determines the nature of an individual's response in a given situation and, what may or may not be the same question, how does an individual select the particular schema(s) to be used.

So, to summarize the essential points I have tried to make so far, there are three things which must be investigated in order to understand mathematical knowledge and its acquisition: problem situations, schemas and responses. One must consider the difference in the problem situation as it is intended by the observer and as it appears to the subject. One must understand the nature of schemas and the means by which they are constructed. Finally, it is necessary to explain how the subject selects the schemas to be used in the response and what determines the kinds of new constructions (if any) that are made.

That represents a very large research program. Some of the issues mentioned have been studied for a long time by a lot of people. Some have hardly been considered. My own research has been concerned with a small part of the program. In observing students in the process of trying to learn mathematics concepts at the late secondary and early post-secondary levels, I have tried to explain their activity by describing specific schemas that they could be using and specific constructions that they could be making. In this connection, I have developed a general description of a schema and an analysis of several means of construction that seem to be available to the individual.

The Constructive Aspect of Reflective Abstraction

In this section I will consider what happens when the subject, in responding to a problem situation, is not satisfied with existing schemas and attempts to construct something new. The means available for construction will be specifics of what I call the *constructive aspect of reflective abstraction.* Thus, my goal in this section will be to explain what I mean by schemas and to describe what seem to me to be the constructions that a subject can make (which will lead to new schemas) in trying to deal with a problem situation.

Schemas

A schema, as I have said, is a more or less coherent collection of cognitive objects and internal processes for manipulating these objects. A schema includes various actions and constructions that can form new schemas. Unfortunately, the only way I know of explaining these various ingredients is to discuss them separately. They are, however, integrated parts of a totality and this is why it is difficult to discuss one of them without mentioning some of the others.

Now, for me, all objects are cognitive. They are constructions which a subject makes in order to deal with various experiences. These experiences can have to do with sensory perceptions, motor activities, or thought. Thus, rock, seven, the cosine function, and the Klein group can all be objects. It may sometimes be useful to distinguish certain kinds of objects (called physical) from other kinds (called mental) but as far as existence goes it is only a taxonomy and, I believe, without any intrinsic distinction. It is perhaps a feature of mathematics that, as the topics become more advanced, the interest becomes more focused on what this taxonomy might call mental objects, and so it is less controversial in this context to say that objects are cognitive. In any case, I will often omit the modifier.

I use the phrase "internal process" (often shortened to "process") as distinct from procedure or action. By the latter I refer to a set recipe of activities which the subject can engage in. A procedure will have a fairly rigid structure which is not necessarily linear, but allows for alternate paths with choices based on the perceived results as the procedure moves along. The important feature of this structure is that it is completely local. That is, aside from the fact that it may be totally committed to memory or even written down, the subject only understands it locally. He or she is aware of how to decide what to do at a given point (go to the next step, chose one of several possible next steps, etc.) only in terms of that point itself, and not in terms of any overall picture. Indeed there will be very little awareness of any kind related to the procedure and the subject will not be able to reason about it. The subject always knows what to do next, but not much else.

An internal process, on the other hand, involves a subject's ability to mentally manipulate cognitive objects. I mean by this that what may have been a procedure as described above, now consists of an individual actively moving mental objects around, calling them into awareness, combining them, comparing them, ignoring them, transforming them, etc. all in her or his mind. The subject is not only aware of the individual steps of the process but has a total picture of it and can move back and forth performing and reversing the mental activities. As a result of her or his awareness of the total process the individual can reflect on the process itself, combine it with other processes, reverse it, reason about it, etc.

It can be difficult to decide whether a particular activity of a subject is a procedure or an internal process. Consider, for example, arithmetic operations on polynomials. You can add two polynomials, multiply them, multiply a polynomial by a number, divide one polynomial by another (obtaining a quotient and a remainder) and substitute a number for the letter in the polynomial to obtain a numerical value. Now, for the high school student who is learning to do all of these things, they can be procedures, fairly devoid of any meaning; for the mature mathematician who is studying the ring of polynomials over the real numbers, these actions are very likely processes; and for the undergraduate mathematics student who has been given the task of constructing a package of computer procedures to implement polynomial arithmetic, they may be something in between. My point is that the distinction between procedure and process is not something intrinsic to the mathematics concept, but has to do with the relationship between a subject and an idea.

Constructions of Processes and Objects

Construction, in this context, is a sprial in the sense that objects are used to construct processes which are then used to construct new objects from which new processes are formed and so on. I am not here concerned with the issue of a starting point (if there is one) but rather will break in at some point, assuming that whatever requisite constructions are necessary have, in fact, been made.

I will consider five means of construction because these are the ones which I have inferred from my observations of students: *interiorization, encapsulation, coordination, reversal and generalization*.

Interiorization. My assumption is that the construction of a process begins with actions on objects which are organized (possibly as a procedure, initially) and interiorized with awareness of a coherent totality. An important example which we will consider below in connection with coordination occurs when an individual constructs a particular function in her or his mind. For instance, a student might read, or otherwise become aware of, a description of a function in the sense of a domain and range (possibly implicit) and a formula. The student may perceive a problem situation which is to make sense out of what is written. If the domain and range sets are understood as sets of objects, then one possible response would be for the student to construct a mental process which uses the formula to manipulate the objects (in her or his mind), and transform elements of the domain into elements of the range.

Another example which I will consider here arises when a person is trying to understand a quantified statement. In a study which is described in more detail elsewhere (Dubinsky, in press *a*; Dubinsky, Elterman, & Gong, 1988) we interviewed a number of students (mostly sophomore computer science majors) who were taking an experimental course in discrete mathematics. In the interview, each student was asked to explain how to determine the truth or falsity of the following assertion.

> For every tire in the library, there is a car in the parking lot such that if the tire fits the car then the car is red.

Here is a sampling of some responses.

FOS: It seems like each of these little pieces, it just iterates through each piece. Like it takes a tire, and then it takes a car and it checks if the tire fits the car and the car is red. And then it takes the next tire-- well, actually it takes the same tire and goes to the next car.

The first student, FOS, appears to have interiorized a process of iterating through the tires and the cars. She does not say anything about the set(s) through which the iteration runs and at first seems to be considering the two variables (tire and car) as a single unit. In the end she differentiates the two but there is no explicit suggestion of any quantifications controlling the iteration.

FAL: Well, all right, I'd go through all the tires that are in the library, and take the tire and take it out to the parking lot, and if it fits, then see if the car is red.
I: What if it fits and the car is red?
FAL: Then it's true.
I: And what if it doesn't fit?
FAL: Okay, if it doesn't fit, then, er ... , then it's still true.
I: So when would it be false?
FAL: If the tire fits and the car isn't red.
I: So, if you find a car for which it's true, what would you do next?
FAL: In order to do what?
I: You said you were doing it for all the tires. So you take a tire, you go to the parking lot, and you said you were going to check all the cars. And if you find one for which the statement is false, what do you do at this point?

FAL: I go on to the next tire.

I: And if for one tire, you've tried all the cars and its false, what do you do?

FAL: Let's see ... I just go on to the next tire.

I: And how would you decide whether you return true or false at the end?

FAL: Well ... it ... I mean. Given any tire, you take a tire out of the library and it says there is supposed to be a car such that if the tire fits it then it's red.

FAL is quite explicit about the domains of the variables and strongly differentiates the two, apparently to the point of ignoring variation in the car. Even when the interviewer tries to lead FAL towards thinking about different cars, he sticks to "the next tire." Again there is no indication of any awareness of the effect of a quantification.

AZU: Run around the library, pick up all the tires, try each tire on a car and see if the car is red. If you find a tire that fits a car in the parking lot and the car is not red, then the statement is false.

AZU seems to apply universal quantification to the iteration over tires, but not cars. Probably as a result, his final conclusion is incorrect. In the next two examples, AOK and GRA apply universal quantification to both variables. Thus they appear to be differentiating the variables but not the quantifications.

AOK: Well, you would iterate through every tire in the library and you would look at every car in the parking lot and see if the tire fits the car and if the car is red. Run through all the set of tires in the library and cars in the parking lot and see if this is true or not.

GRA: The best way would be to go in order, first finding every tire, every tire in the library, and finding all the cars in the parking lot, and for each tire, taking physically, taking every tire and seeing if it fit the car and checking if then the car was red, and if for each of those tires the car was red, if it fit, that would be true.

Next, we see that VLA describes a much more structured procedure. The fact that he is able to go back and forth, correcting and revising, until he finally ends up with a description that satisfies him, suggests that he is working with an interiorized process. Because of the sloppiness of the language (unlike the previous students, he is not a native speaker of English) it is a little difficult to compare his process with the two-level mixed quantification that I construct as my understanding of this statement, but it is reasonable to infer that they are not incompatible.

VLA: I have a set of library tires and then I would have to check one at a time, pick a tire out of the set of library tires, then go over to the parking lot and pick one car and then check ... then for ... I would take the first tire out of the set of tires and the first car and check that if the tire fits on the car then the color of the car is red--and I would have to repeat this--I would have to ... uh ... find I would have to go through this procedure on all the cars in the lot until I find one that fits and then I would stop because it doesn't matter and then I would have to repeat this procedure for all tires.

Finally, the response of ELK (also not a native speaker of English) is about the same as what most mathematicians would say and thus suggests that what ELk has constructed is not inconsistent with conventional interpretations of the statement.

ELK: Well, let's see. One way to do it would be to look at every tire in the library and then go out and find at least one car in the parking lot so that if this tire fits the car and check to see if the car is red.

In each case, my inference is that the student has constructed an internal process as her or his response to the question and the verbal statement is an attempt to communicate this process. In working further with this statement, responding to additional questions that were asked, I assume that *amongst other possibilities* the student could use this process to reason about the statement.

Encapsulation. Encapsulation, like interiorization, is particularly difficult to see directly in any sense and can only be inferred. That is, one can only observe subjects who seem to be able to think only about a process in a particular situation and other subjects who, in what appears to the observer to be essentially the same situation, seem to be considering an object. The inference that is suggested by the theory I am putting forth is that the latter subjects have encapsulated the process into an object.

In a study (Dubinsky & Lewin, 1986) that took place earlier than the one on quantification, we asked students to explain what they meant by "proof by induction," concentrating not on the formal definition, but in trying to get at the student's understanding of why (or if), once such a proof was made, the statement was true--and what that meant.

We found many students who had difficulties at an earlier stage, even before the issue of proof arose. For example, in the following response, STU refers to the process of moving from one value to the next by adding 1, but does not refer spontaneously to the implication between the two propositions obtained by evaluating the statement at two successive values of n, even when prompted with the word. He is, however, able to pick up the implication process when it is mentioned by the interviewer at the end.

STU: You're given an equation for n ... you prove it for the first n, you put the 2 in there and you prove that's true. Now you want to prove that it's true for each subsequent n ... and you prove that's true.

I: When you prove it's true for n + 1, what is the thing that you use? What is the most important thing that you use?

STU: The equation.

I: But why did you prove it then for the first one?

STU: You prove it true for the first value.

I: Right. And then what do you do?

STU: Add 1 to that, prove that's true.

I: Isn't there an implication in all that?

STU: That's true for all n.

I: But isn't there a kind of next step implication?

No response

I: How would you prove it true for 4? You've got it proved for 3, right? Now how would you prove it for 4?

STU: I just always thought that once you prove it true for one number, and 2, 3, 4, 5 and proved it true for n + 1, that it was true for all the numbers.

pause

Just because you prove it true for the first statement, n = 2, just means it's true for 2. It doesn't mean for any other positive integer.

I: That's right. If you prove it true for 2, then it means that it's true for 2.

STU: Okay, so what am I missing?

I: What you're missing here is that the step here is that if it's true for n, then its true for n + 1.

STU: I think I knew that if you prove it's true for n + 1 and n + 1 would bring it to 3, and now if it's true for 3 and it's true for n + 1, that would bring you to the 4, if its true for 4, and it would be true for the n + 1, which brings you to 5. I couldn't say that, though, I suppose.

On the other hand, when KAR refers to "the same thing," in the next protocol it appears that she definitely has an interiorized process for using implication to go to the next step, and is aware of it. Her response after the pause comes when she is asked to explain, once the proof has been made, why the statement is true for n = 10. Her use of the phrase, "same thing" suggests an awareness of the implication as a totality, which under prompting, she can see as a varying object, which goes beyond what STU was able to say.

KAR: First of all, you prove that something is true for 1, for n = 1, and then it's true for n, then if it's true for n + 1, then its true for all n because that's all the natural numbers.

pause

It's true for 1, and it's true for n + 1, and 9 + 1 is 10, and 9 is a natural number, and its true for 9, so its true for 10.

I: How do you know it's true for 9?

KAR: I could say the same thing with 8.

I: Say it.

KAR: Okay, it's true for 1 and it's true for 8 so it's true for 8 + 1, which is n + 1, so its true for 9.

And in the following exchange, SUS, in referring to a pattern, seems to be talking about a total object.

SUS: If you want to show it for a million, you say its equal for 1, then you go on and say it's equal for 2 and then you go on and say its equal for 3, and so forth.

I: Right.

SUS: You're starting to develop a pattern, once you get that pattern going, then you can kind of see ahead, and even if you go further than a million, you can see that n would still be an element of that expression.

We can consider another example from the quantification study (Dubinsky, et al., 1988) referred to above. The following responses came when the students were asked to explain how to determine the truth or falsity of the following statement.

> Amongst all the fish flying around the gymnasium, there is one for which there is, in every computer science class, a physics major who knows how much the fish weighs.

I am concerned with the process (which depends on a fish that has been chosen) of determining that every computer science class has a physics major who knows how much the fish weighs. Here are the responses of two students who were asked about this.

REI: Okay, you take all this--you would take the set of fish that are in the gym and set of students that are in Computer Science--you would take the set of all the computer science classes and the set of all the students that are in those Computer Sciences classes and check to see if there was a student in one of those--in every one of those--yeah, at least one student in every one of the classes that knew how much one of those fish in the gym weighed. And if that were true, you would return true and if it found one case where that failed, it would be--if there was one class with no students ...

I: Okay. Would you do it exactly the same as ...

REI: (Pause) I'd probably go class by class and ask in each class if there was somebody who didn't know how much any of the fish in the gym weighed.

I: So you would have skipped the first set--you skip the set of fishes?

REI: Yeah. I think I'd probably try and prove it false, rather than trying to prove it true.

and

VLA: Okay, I would look at a set of fish among the set of all available fishes and I would have to iterate over that and of course the condition is that as soon as I find the first one for which the rest of the long expression holds, I stop right then and there.

I: Can you explain to me what would be the rest of the whole expression? How would you check that?

VLA: Yea, that was just the first step.

I: Good.

VLA: I got ... I'm picking a fish and then I have to start iterating over a set of available classes. Here I'll have to go through every one of them for that fish. And then I would have to go through a set of students in the class. Here we're dealing with an exists so that as soon as we find the first one that matches the rest of the conditions, its fine. And then I would run that function on the student.

I: What would you ask about the student?

VLA: I would ask if the student knows the weight of the fish.

It would seem that REI has this process but, unlike VLA, is not clear about its dependence on the fish. In his reference to "the rest of the long expression," and "go through every one of them for that fish," it could be that the two-level quantification over classes and students is a totality for VLA and this object depends on the choice of the fish. Our inference is that both students have constructed an internal process to deal with this two-level quantification and that VLA has encapsulated the process into an object, but REI has not.

In informal observations of students trying to learn mathematics, I have seen many examples of what seems to me to be the issue of encapsulating a process. If you describe a set which contain other sets as elements such as

$$\{7, \{6.3, -5\}, -2\}$$

and ask students how many elements this set has, many students will say, 4, and will be quite surprised when they hear the interpretation that it has three elements: two integers and a set. The issue here is not about which answer is right and which is wrong, but concerns the student's difficulty in seeing that the set $\{6.3, -5\}$ is itself an object which can be an element of a set.

My inference about sets is that, at one point, the concept of a set is a process consisting in something like putting objects in an imaginary container, or a process of deciding whether an object is or is not an element of a set. A next step is to encapsulate this process so that a set is an object which can be in another set, vary with a parameter, etc.

Still, another example is the concept of function. It is important for many operations with and uses of functions that the subject construct processes as her or his understanding of particular functions, as described in the section on interiorization. Later, however, when one wants to consider sets of functions, parameterized families of functions, functions whose domain and/or range is a set of functions, etc. it seems essential to be able to treat a function as a total object. We will consider this further below in the section on functions.

Coordination. Returning now to processes, we consider the cognitive act of taking two or more processes and using them to construct a new process. This can be a simple concatenation (composition in the case of function), a simultaneous consideration of several processes, perhaps organized in nested loops, or something much more complicated as in the case of induction (see the section on mathematical induction) where it is necessary to coordinate the process of an implication-valued function of the positive integers with the process of modus ponens.

In my experience, interiorization of a process is usually difficult for students and encapsulation is always troublesome. Neither seems to happen in one fell swoop nor is a simple suggestion enough. They both require some sort of influence that goes beyond verbal explanations. I have found, however, that once a student has interiorized the processes involved, coordinating them seems relatively natural and can appear to occur in response to a suggestion or explanation--even in a lecture. Remember, however, that I am talking about students at least at the secondary level and situations in which the problem of being aware of more than one thing at the same time is not an issue.

The composition of functions is a particularly informative topic to study in the context of internal processes and their coordination. In the same class that was subjected to the quantification study, I conducted some interviews after using teaching methods based on my analysis of the concept. The student LIE who made the following response is fairly bright but not particularly interested in mathematics and shows no unusual talent for the subject. I think his understanding of the function process and his aiblity to coordinate processes are fairly strong. In reading his descriptions, one can follow what he says by writing symbols on paper and the result is an excellent mathematical analysis of the task. In such an example, I feel that one can almost see the constructions in the reflective abstractions taking place before one's eyes.

LIE was given a sheet of paper with the following two statements on it.

1. Take a triple of integers and return the value of the sum of the second and third integer.
2. Take a triple of integers and replace the second one by the sum of all three intergers. The first and third are unchanged.

He was allowed time to write whenever he wanted during the interview, and, in the transcript, he often is reading from what he has written.

I: On this sheet of paper are two statements. Can you explain how they define two functions.

LIE: Well, let's see, the first one is a function and it takes in three inputs and ignores one of them, and adds the other two. It takes in three things and outputs the sum of the last two things it took in. And then for 1...so that for 2, it takes in three integers and replaces it, outputs the first integer unchanged, the sum of all three integers, and the third integer unchanged, the sum of all three integers, and the third integer unchanged.

I: You think so?

LIE: Well let's see. If the input, x1 comma x2 comma x3 equals x1 comma plus x2 or maybe there should be square brackets. So it takes in those three numbers and it outputs those three (pointing to what he has written on paper.)

I: Now suppose we call the first function *F* and the second function *G*. Can you find a function *H* such that *F* equals the composition of *H* with *G*? So *H* of *G*.

LIE: Well, o.k., this is, what I'm trying to do is get what I wants is *F* of x1 x2 x3 equals the value of the sum of the second and third, equals x2 plus x3 and so that has to equal *H* of *G* of x1 comma x2 comma x3, and that has equal *H* of x1 comma x1 plus x2 plus x3, comma x3. So x2 equals b minus c minus a, x3, so you do *G* first, x1 plus x2, *H* of x1 comma x2 comma x3 equals x2 minus x1.

I: OK, can you explain how you got that?

LIE: Well, the first thing I did, I wrote down that *F* equals *H* of *G*. So you want *F* of x1 x1 x3 to be equal to x2 plus x3, so that means that you want *H* of *G* x1 x2 x3 to be equal to x2 plus x3. Well since we do *G* first because that's *H* of *G*, that means we know what *G* of x1 x2 x3 is, its x1 comma x1 plus x2 plus x3 comma x3. So that means that x2 plus x3 has to equal *H* of x1 comma x1 plus x3 comma x3. So then I just give different names, single names, to each of the three inputs of *H*, call them a, b, and c. Then I realize that x2 is equal to b - c - a and that x3 was equal to c so x2 plus x3 is equal to b - a so that means that each of a, b, c is just b - a and I just renamed a, b, c to be x1, x2, x3, and wrote down *H* of x1 x2 x3 equals x2 - x1.

Reversal. It should come as no surprise to anyone familiar with Piagetian theory that reversal plays an important role. It arises in an interesting way in connection with the above interview of LIE. The problem he was given was to solve the following equation

$$F = H \circ G$$

for *H*, given *F* and *G*. It turns out that the problem would have been much harder if he had been given *F* and *H* and asked to solve the same equation for *G*. The difference in difficulty is important because of the fact that, mathematically, the issues in the two problems are identical. I have asked many mathematicians about their relative difficulty and I can find no leaning towards either. An analysis of these two problems, evidence on their difficulty, and a disucssion of the connection with reversal is given in Dubinsky (1990) and discussed below in the section on functions. Here we consider a different situation, the concept of functions that are 1-1 and/or onto.

Students who had been using computer experiences to study the function process and its relation to the ordered pairs idea of a function were asked to explain when a set of ordered pairs represents a function and to explain what is meant by 1-1 and onto, both from the general point of view for functions and how to check these properties when the function is given as a set of ordered pairs. Also, they were given two specific examples of functions (presented as sets of ordered pairs) and were asked to determine if they were 1-1 and onto. Almost all of the students, including the ones whose interviews are excerpted here, gave answers that were considered to be correct *except* for the general explanations of 1-1. That is, their answers to all of the questions about *onto* functions agreed with normal useage and for 1-1 they were able to decide about specific examples, but their general definitions did not agree with the standard definition and if applied to the examples, would have given answers different from what the student gave. In other words, in deciding about examples, the students appeared to use the "correct" definition, but in explaining what 1-1 means in general, they used a materially different definition.

In trying to interpret this result, I suggest that both the property of being a function and of being onto requires thinking about the function process in its original direction. That is, one can express the idea of function by beginning with an object in the domain, think about moving over to the range and assert that one goes to a unique object in the range. For onto, one can think about doing this for everything in the domain and *then* moving over to the range and seeing if everything was reached. To think about the idea of 1-1, however, it seems to me that one must again apply the function process to everything in the domain, then move over (in one's mind) to the range and, considering each object there, *look back to the domain* and see if it "came from" more than one object in the domain. In this latter step, I suggest that there is a process of reversal that is not present for the first two and I offer this as an explanation of the added difficulty.

Indeed, if, in a proposed definition of 1-1, everything was correct except that the reversal was not made, then the result would be exactly the criterion for a set of ordered pairs to represent a function. This was, in fact, the error that most students made.

Now let us look at some interview excerpts and see if we can see these processes and their reversals (or lack thereof). In the first excerpt, S1 attempts to respond with a static, formal statement about ordered pairs and spontaneously begins to speak about a process, only after the interviewer indicates that there was something wrong with her response. She appears to be unable to reverse the process, however, even with considerable prompting, and her definition is still that of function rather than 1-1. This is typical of many responses of the students. The resistance of S1 is particularly strong and, knowing that her definition has to be different from that of function, she introduces the 1-1 criteria, but includes it with the criteria for function. An important point is that she reverts to the static definition in order to do this so that it is possible to infer that she did not reverse the process. In the end she reverts to her original criteria and responds to the last question, which is a case in which it would not be a functin but could be 1-1, by saying that it is not 1-1.

I: What does it mean for a function to be 1-1?
S1: That there is ... 1-1, that there is one, if each first element of, if the ordered, the first element of each ordered pair only appears once, its 1-1.
I: But I thought you said that you needed only to appear once for it to be a function.
S1: Okay, so 1-1 is when you've got one, each first ordered, each first element only maps to one second element so that each, you're only going to get each of them once.
I: The each of what once?
S1: One member of the domain will map to one member of the range and you'll never get two of each.
I: Good. If you have a set of ordered pairs, how do you determine that it represents a 1-1 function?
S1: You can, well if anything appears twice in the first or twice in the second, its not going to be 1-1.
I: Well, if it appears twice in the first, what would you say?
S1: It's not 1-1.

The next student is more clear about a process ("...is mapped to..."), but is not completely sure and in fact expresses both the function criterion and the criterion for 1-1 in terms of a forward process. He then confuses these with onto. Of course one could argue, both with respect to S1 and S2 that they have the concepts but simply have the wrong names attached to them. Perhaps, but in all of the examples (which were a little complicated) they got every answer correct with the right "labels." Thus I infer that they were at least unsure about the reversal of the process.

I: What does it mean for a function to be 1-1?
S2: Um, that each, each uh, each value in the domain has only one in the range. Has only, yeah, has only one other value as a second element.

I mean, let's see, every time you find, each number is only used once as a first element.

I: Okay, so how do you determine that a given set of ordered pairs represents a 1-1 function?

S2: If two of the things, if two, alright, there's two ways to do that. If one element's mapped to two different things, or if two different elements are mapped to the same thing. Wait a minute, I think I'm getting onto ...

Finally, here is a student who gives a definition that is considered correct.

I: Now in general, what does it mean for a function to be 1-1?

S3: Every...

I: A general function.

S3: Its for, its when for an element in the range, (*short hesitation*) there's only one element in the domain that will give that element.

It is perhaps not unreasonable to infer that S3 ran the process forward in his mind and that the hesitation was a silence while he was mentally reversing it, and then he looked back to the domain and finally stated the condition.

Generalization. The last construction that we will discuss is generalization. By this we simply mean that an existing schema is represented and used in a new situation different from previous uses. If one considers the schema by itself, then one could argue that it has not really been changed. But my point of view is to coordinate the schema and its use. Thus, a new use of an old schema is, for me, a new schema.

A good example is the function schema.

The students that we interviewed about induction in connection with encapsulation were taking a course in advanced calculus. They were able to use their function schema to interiorize procedures as processes when they were considering functions that transformed numbers to numbers. The situation with induction, however, involves what a mathematician might refer to as a boolean valued function of the positive integers. A generalization of the function schema in this case would consist of the student interpreting a statement to be proved by induction as a function which transforms each positive integer into a proposition which has the value *true* or *false*. Of course a precondition for this generalization would be that propositions are objects for the student. In many cases, a proposition is a process, of determining its truth values, or combining several statements with logical connectives, or going from the hypothesis to the conclusion in an implication, etc. It is necessary to encapsulate these processes so that they are objects which can form the range set of a function.

The following excerpt suggests that S4 is not generalizing his function schema to the extent of thinking about transforming each single positive integer into a proposition.

S4: When you're trying to prove something, you would start off first by proving something that's either obvious with one element and you're going on making it a little more geeneral, and then work yourself into a position where its got to be true for whatever you're proving.

I: In induction usually you want to prove certain statements true for every integer n. So in terms of those integers what would you do?

S4: First prove it for one integer by proving it directly.

I: All right, good.

S4: Then it would depend a lot on what you were tyring to prove, I guess.

I: Well the overall format of proof does it. The overall format of proof is the same every time, but what you actually do to implement that format may change. What I'm more interested in today is, in the extent to which you understand the overall format. I don't care about the details of actually putting it into practice.

S4: Then I guess you would prove it for one directly, then you might prove it for a set of integers, or a group of them, I guess. Between this and this, make it, and then I guess doing theorems and postulates you have, you try to make it more general.

S5, on the other hand uses phrases such as "valid for one", "show it for all n", "equal for one", etc. which suggest that for her, the statement to be proved is a process that converts integers into something that must be shown to be true or false.

S5: You have to show that it's valid for one, they you show that if you have a special n in it, that that would work for that expression and you try to go for the next n, so that's an element of the expression, and then the next one. In other words you try to show it for all $n+1$ that's in the expression.

I: Well that's right and convinces me and so forth, but suppose you were talking to people who didn't know very much about this thing and suppose you wanted to explain that to them, that that's how you prove it for all n, and they said, "But I don't understand." How does what you did guarantee that it's true, for instance for n equal to a million? How do we know that?

S5: If you want to show it for a million, you'd say, if it is equal for 1, then you go on and say, it's equal for 2, then you go on and say it's equal for 3, and so forth.

Recapitulation of the Construction of Schemas. To summarize, we can say that a schema consists of cognitive objects and internal processes. It includes the possibility of performing actions on these objects. A collection of actions can be interiorized to become a process and this is one way that processes can be constructed. Another way is to coordinate two or more processes to obtain a new process. Also, a process can be reversed which is still another way of constructing a new process. On the other hand, when the subject has a high degree of awareness of a process in its totality, this process can be encapsulated to obtain an object. This is the way that objects are constructed. Finally, a schema can be generalized by applying it in a new situation to which it has not previously been applied. The notion of schema is described schematically in Figure 1.

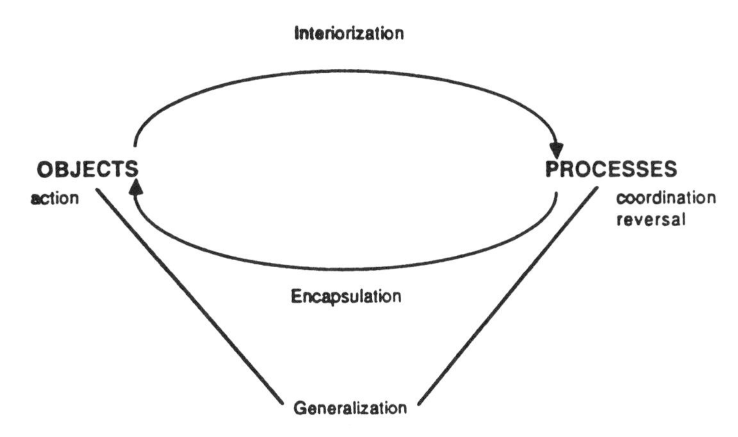

Figure 1. Schemas and Their Construction

My Research Program

As I said earlier at the end of Section 1, what I am proposing is a very large research program and must concern the problem situation, the nature of schemas and responses. At the present time I am working with a small part of it that concerns the last two items. I am trying to understand how the general theory relates to specific topics. That is, I am considering individual topics and trying to work out what I call *genetic decompositions* for them. A genetic decomposition for a topic is a detailed description of what a schema for this topic might look like and how the constructions described in the section on construction of processes and objects could be used to construct this schema. In addition, I am trying to develop teaching methods that are compatible with the genetic decompositions, but these will not be described here.

The Paradigm

The paradigm has three steps: analysis of the concept, design of instruction, and observation of students. One can think of these as taking place linearly in time, but in reality, each step relies on the other two and all three are repeated with, hopefully, increased understanding and effectiveness.

Analysis of of concept. Because the goal of my research includes both an understanding of mathematical knowledge and a development of teaching methods for helping students acquire it, I believe that it is necessary to construct, in my own mind, a description of any concept that I would like students to learn. I call this a genetic decomposition of the concept and I consider it to be the same as what von Glasersfeld has referred to as "an analytical model of the adult conceptualizations towards which his (the teacher's) guidance is to lead" (parenthesis added) (von Glasersfeld, 1983).

What is the source of this model? I believe that although my own understanding of the concept in question (to the extent that, through introspection, I can become aware of it) is an essential ingredient, and indeed it would seem impossible to keep it from having a major effect, there are two other sources which must be brought into the analysis. One is the general theory that I have been describing in this paper, and the other is the observations of students in the process of trying to understand the concept.

It would be a very serious error to allow the researcher's understanding of the concept to have an undue influence. There are many examples in the literature and in my own work which suggest that the understanding that a mathematician (who may be unusual) has of a concept, long after it is learned and has been applied, is often very different from what a typical student may be working with. The researcher's understanding is perhaps most important in the beginning, before any observations have been made, when *something* must be used to decide what instruction to begin with, what questions to ask, what tasks to be set. The point is that, if we are talking about more advanced mathematical topics, such as say, proof by induction, one cannot simply set a student in the situation. One must create a fairly complex context before it is even possible to observe anything relative to the concept. As the investigation proceeds, the role of the researcher's understanding of the concept should decrease sharply although it can never be eliminated.

Similarly, the design of initial activities must be strongly influenced by the researcher's theoretical background which, for me, is the theory described in the paper. As the investigation proceeds, the influence of that theory does not decrease so much as reverse direction. That is, as the students are observed, the theory is used to analyze the observations, but it is a two-way street in the sense that it often happens that important revisions of the theory result from the observations.

I cannot emphasize too strongly that the analysis of the concept presents a very serious danger. A distortion of my approach could lead to powerful activities that would attempt to fit students to an adult mold. The researcher must make a special effort to avoid becoming too attached to her or his analysis. The researcher must understand the tentative nature of the theory, and the strong liklihood that it will be changing. It and the researcher's understanding of the concept must be relegated to the role of initial guide and not be allowed to exert an undue influence.

Design of instruction. The instruction, as in all cases, must take into account various affective issues. Student beliefs about mathematics and about their ability to learn it, motivation, the exigencies of the real situation (class size, institutional concern with grades, etc.) are all matters which cannot be ignored in any design of instruction. What I am concerned with here is in addition to these issues. Again, however, although I will discuss my concerns as if they could be isolated from the others, this is not really the case and in actual teaching situations, one must coordinate everything. One particular consequence of the impossibility of isolating issues is that evaluation becomes a very cloudy business. Fortunately, I am allowed to ignore evaluation in this paper--if not in the research.

There are two important points relative to instructional design that I am concerned with here: getting students to be aware of what they are doing when they perform mathematical tasks and inducing them to make the appropriate constructions of the types described in the section on construction of processes and objects. That is, when designing instruction, I don't ask myself questions like, how can I explain this point, or even how can I get them to understand this idea. I ask: how can I get students to interiorize a process which corresponds, for them, to what for me is a particular function; how can I get them to coordinate two of these processes in response to situations in which they might need to construct the composition of two functions; how can I get them to encapsulate their various processes for various functions, so that we can talk about sets of functions, or parameterized families of functions?

Other than the fact that I include computer activities in the tasks that I set for students, I don't have much in the way of a general answer to these questions. At each point in a course, I ask myself these questions, and all that I have is a large collection of examples and some idea (a combination of opinions and observations of student performances) about what seems to work and what doesn't.

What I can say in general is that my instructional activities do not include very much lecturing. Mostly, class time is spent in doing tasks and talking about them. Occasionally I will explain something, but mostly I consider my activities to be indirect and that the main issue is the constructions that the students may or may not be making.

Observation of students. I often ask students to perform tasks in writing or on the computer and to give me a record of what they did. These tasks have two purposes. The first is to present them with a situation which I hope they will perceive as a certain kind of problem situation. My assumption, as I have said, is that this will result in constructions that will add to their understanding of the mathematics concept that I am concerned with. The second is to find out something about the constructions that they have or have not made.

The more I take written records from students and attempt to look further into what is or is not understood in individual cases, the more I become convinced that a single response, by itself, is rarely conclusive. A student may know the answer to a question, but not give it; he or she may have constructed a schema but not use it; it is not only possible, but common for students to respond in very different ways to what seem to me to be several instances of the same situation; I have often had the experience of asking a student a very difficult question, getting a response that I considered far from adequate, asking the student to try again, and getting a perfectly reasonable answer. For this reason I believe that it is essential to go beyond the single, written response.

The clinical interview, in which the observer has a strong idea of what he or she would like to get at, but is completely open to the student's idea of how to get there, is an important approach. These interviews are recorded and subsequently analyzed.

Recently, I have used a combination of written response and the clinical interview. In this way I can set a task for a large number of students and, from the responses, select a small number of individuals for a more extensive interaction in an interview situation. It turns out that although student responses can be very different, the variation is often much smaller than the number of students.

The revision of the genetic decomposition, which is repeated as long as the investigations are continued, is one of the most important aspects of the research paradigm. The key ingredient in the revisions is the set of observations of the students. In analyzing these protocols I look for comments that suggest to me that certain reflective abstractions have taken place and certain constructions of the type described in the section on constructions of processes and objects have been made. Alternatively, when a student does not succeed with a task, or is confused, I try to see if making a particular construction would have clarified the situation. When this occurs to me during an interview, I prompt the student to make the construction.

Revisions of the theory occur when I see what I consider to be constructions of a type different from what I have noticed before.

Summary of the paradigm. So the circle runs as follows. I begin with a genetic decomposition of the concept (initially from my own knowledge, guided by the general theory), I design instruction intended to induce students to perform the constructions which it specifies, and then I observe the students as they try to make sense out of this instruction. Based on my observations, I revise the genetic decomposition, and possibly the theory, and repeat the process.

How the Paradigm is Developing in My Own Mind

I would like to share with you how this research paradigm has been growing in my own mind. It is today very different from what it was at first, and I expect that in the future it will be very different from what it is today.

It all begain about six years ago when I was teaching a discrete mathematics course at a university which required of their students a very extensive involvement with computers. It was a time when computer science was extremely popular with undergraduates. I asked the computer scientists at my school what they would expect students to get out of my course and their answer, which surprised me, was that the most important thing was that the students should learn to think abstractly. The computer scientists considered that abstract thinking was essential to their discipline and that students were not very good at it.

In trying to meet these expectations I noticed that students who had been working with computers seemed to be a bit more amenable to certain mathematical notions, than other students I had worked with. For example, an axiomatic system such as a group could be, and was considered (not spontaneously, but at my suggestion) as a kind of a programming language.

It was also the case that certain computations, such as figuring out all groups with 1, 2, 3, 4, and 5 elements, could be used to stimulate conceptual development.

I wanted to pursue the ideas of computer analogy and computation so, in teaching an advanced calculus course, I tried to get across the idea of induction by using "while loops" in Pascal. It did not seem to help at all. Because I was beginning to study Piaget at the time, I decided to interview the students about induction. Some excerpts of these interviews are given above in the section on the construction of processes and objects. What I saw there was profoundly disturbing. Not only did the students (junior and senior mathematics majors) seem to have very little understanding of induction, but it seemed clear to me that even the very simplest constructions had not been made. I began to realize how little value there was in the usual way in which we "explain concepts" to students.

There then followed several years in which I thought about a general theory at the same time that I conducted observations with induction, sets, propositional calculus, predicate calculus, and now, functions. The overall paradigm as described above in the summary of the paradigm came fairly early as a derivative of Piaget's ideas. Initially, I was under the impression that the genetic decomposition came entirely from observations of the students. Gradually, as I saw that there were aspects of the genetic decompositions that could not be found in the interviews, I realized that my own understanding of the concept must, and should play a role. As the theory developed, I consciously applied it in my constructions of genetic decompositions.

The list of constructions has also grown in time. In first working with induction, I only saw encapsulation, coordination and generalization. These were not enough to explain what I was observing as students struggled with predicate calculus and I realized that interiorization was taking place. Looking again at induction, I saw that including interiorization improved the explanation. This was reinforced by my growing realization of the importance of interiorization in Piaget's theory and it is proving indispensable in my thinking about functions. It is only recently that I have begun to see reversals of processes and at the moment, it is kind of tacked on to the theory. Presumably, future developments will suggest a more integrated role.

Finally, it is a recent development for me to organize things with objects and processes playing the central role. This has clarified matters for me and it also relates to the issue of the nature of the reality of objects. Many people have difficulty with the notion that objects exist in the mind of an individual as a result of constructions which the individual makes. Mathematicians should have less of a problem. For me, the category of topological spaces is an object in exactly the same sense as a particular book.

Each is made of individual components (spaces or pages) but has a totality that establishes it as an object. Now whether you think of the categories that Sammy Eilenberg and Saunders MacLane constructed some 50 years ago, or the categories of analogies that Piaget seemed to believe that each child constructs, it is easy to argue that these objects must be constructed. So, if I can accept that this is true of the object, category of topological spaces, and if I try to remember some of my early ideas of books, it does not seem too hard to accept that I also constructed my concept of a particular book. It does require practice, however.

Specific Mathematics Topics

In this section I will try to show how the ideas in this chapter can be applied to attempt to understand how individual mathematical concepts might be organized in the mind of a subject. First I will discuss topics which I have studied more or less intensively and then I will consider briefly a few topics which I have not studied at all.

Topics That Have Been Studied

I will discuss three topics: mathematical induction, predicate calculus, and functions. I have studied the first two rather extensively and I will present a genetic decomposition which has been derived and revised making use of the full paradigm. Because of space considerations, I will not give any examples of protocols which led to various points in the genetic decompositions. These examples are presented in detail in the papers I referred to and, for the most part, are represented by the protocols discussed in the section on constructions of processes and objects. In the case of functions, I have only made some preliminary investigations which have been used to some extent in developing the ideas which I will present, but it should be understood that they come mainly from my own understanding of the concept of funtion and a conscious attempt to apply the theory that I have presented in this chapter.

Mathematical induction. First of all, let me say some things about induction itself. It is important to go beyond the usual examples of discovering and proving formulas for sums of series (such as the sum of the first n positive integers). There is a rich variety of problems connected with the idea of a statement depending on an integer variable and the determination that the statement is true for all (or nearly all) values of the integer.

Actually there are two kinds of problems. One is to discover the statement and the other is to understand why it is true. There are some problem situations in which both issues are relevant and others in which only the discovery has to do with basic mathematical reasoning. But there are also problem situations in which the statement is obvious, but its truth is not. In my work I have concentrated on the latter category with problem situations such as the representation of dollars with chips and the divisibility of an integer with repeated digits, both of which are discussed below.

In this kind of situation, it seems to me that the issue is the understanding of the idea of a statement being true for *infinitely many* values of a variable and of the notion of establishing this infinite totality of "facts" with a finite process that requires only a starting point and the facility for always being able to "take the next step." Although I have not studied the concept of infinity, it would not surprise me if this view of induction was a part of the early steps in constructing a concept of infinity.

In any case, I will now proceed to discuss the concept of proof by induction as a method for establishing that a given (or discovered) statement depending on an integer parameter is true for (nearly) all values of the integer.

Mathematical induction needs to become an object in the subject's general schema for proofs. Of course proof by induction is a process, but it must have been encapsulated in order that the subject can reflect on it, along with other methods, when confronted with a theorem to prove, so as to select induction as the method for a particular problem.

The method itself is constructed, in the genetic decomposition that I will describe, by working with two major schemas: function and logic. The developments of these two schemas are interwined through various coordinations. We can illustrate the process with a chart as shown in Figure 2.

We start with the assumption that the subject possesses a function schema and a logic schema that are already developed to the point where, for example, the function schema includes the ability to construct a process corresponding to a given description of a transformation of numbers in the section on functions, and the logic schema can construct statements in the first order propositional calculus in the section on the predict calculus. In particular we assume that the function schema includes the process of evaluation of a funtion for a given value in its domain and that the logic schema includes a process for logical necessity, that is, in certain situations, the subject will understand that if A is true then *of necessity* B will be true.

I consider that the function of first order propositions is a process in the logic schema which comes from interiorizing actions (conjunctions, disjunctions, implicaitons, negations) on declarative statements (objects). The subject must encapsulate this process to obtain new objects which are the propositions of the first order propositional calculus, on which the same actions can be performed. Consider for example, a simple proposition such

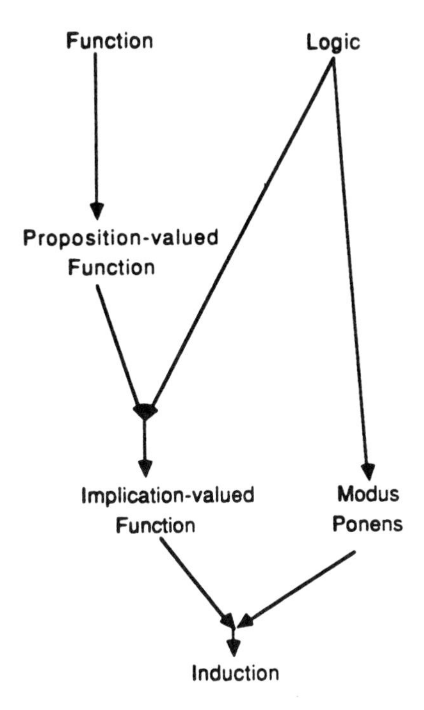

Figure 2. Genetic Decomposition of Mathematical Induction

as,

$$(P \vee Q) \wedge R$$

where P, Q, and R are simple declarations. The formation of the disjunction $P \vee Q$ is an action on the statements P, Q. It is not just the action of putting these symbols in this expression. The subject must also construct a mental image involving the two statements and the determination of the truth or falsity of the disjunction in various situations. If nothing further is done after this action is interiorized, then it will not be possible to combine this with R to get the full proposition. First, the disjunction process must be encapsulated to form a new object $(P \vee Q)$ which is a statement that can be conjoined with another statement, such as R. Note how the use of parentheses in mathematical notation corresponds to encapsulation.

Iterating this procedure, the subject enriches her or his logic schema to obtain a host of new objects consisting of first order propositions of arbitrary complexity. Next the function schema comes in. We are assuming that this schema can be used by the subject to construct processes that transform numbers (for example an integer) into other numbers. It must be generalized to premit the subject to construct processes that transform positive integers into propositions, to obtain what we shall call *a proposition valued function of the positive integers.* Consider for example, a statement such as

> Given a number of dollars, it is possible to represent it with \$3 chips and \$5 chips.

For such a statement, the subject may construct a process whereby, for each positive integer n, a search is made and it is determined whether it is possible to find non-negative integers k, j such that

$$n = 3j + 5k$$

It is useful for the student to discover that the *value of this function is false for* $n = 1, 2, 4, 7$ and then appears to be true for higher values.

It is only at this point that the subject can realize that the problem of "proving" the statement consists of determining that the value of the function is true for all values of $n > 8$. For this, the proof schema can be invoked. If it contains the schema for induction, it can be used, if not, further (re-) construction must take place. In describing this construction, we reiterate the point that, in the context of this theory, it is never clear (nor can it be) whether one is talking about a schema that is present or a schema that is being (re-)constructed.

Before going on with the description, there is a side issue that should be pointed out. Whether the subject is able to construct a proposition valued function of the positive integers as part of her or his understanding of a particular statement depends not only on the existence of the schemas we are talking about, but also may require additional knowledge about the particular situation--so called "domain knowledge." Thus, although the above example of chips is probably well within the domain knowledge of most students who find themselves trying to learn induction, others may not be. We have found, for example, that the following statement provides difficulty for university undergraduates.

> An integer consisting of 3^n identical digits is divisible by 3^n

The trouble seems to lie in understanding the relationship between the value of an integer and its representation with digits. It is a sort of "grown-up" version of the difficulty with the concept of place value and it suggests that many students have not really acquired this concept--at least in a sufficiently powerful form. Returning now to the construction of proof by induction, the next development provides an example of a cognitive step which our research has pointed out as providing a serious difficulty, whereas if one takes only the mathematical point of view, there is not even a step that needs to be taken. This is the case even though it relates to an overt difficulty encountered by everyone who has tried to teach mathematical induction.

We are referring to the notion that the essential point in an induction proof is that one does not prove the original statement directly, but rather the implication between two statements derived from it. This is a major difficulty for students. It requires a cognitive step which is not necessary as a mathematical step. To explain, let us denote by P the proposition valued function to be proved. Now $P(n)$ can be any proposition, in particular, it can be an implication. Therefore, if we define the proposition valued function Q by

$$Q(n) = (P(n) \Rightarrow P(n+1))$$

then, *from a mathematical point of view* there is nothing new in Q which is also a proposition valued function--in which the proposition happens always to be an implication. Once one understand P then, as a special case, one understands Q. We have observed that, with students, this is not the case from the *cognitive point of view.* In the first place, implications are the most difficult propositions for students and they are generally the last to be encapsulated. Furthermore, there is a difference between constructing P from a given statement and constructing Q *from* P. This is the step which must be taken. If there is some subtlety here, then it might help explain the difficulty that students have at precisely this point.

To summarize, this step requires the encapsulation of the process of implication so that an implication is an object and can be in the range of a function, the generalization of the function schema to include implication valued functions, and the interiorization of the process of going from a proposition valued function of the positive integers to its corresponding implication valued function.

The next step is to add to the logic schema a process which we shall call *modus ponens*. This process is an interiorization of an action applied to implications (assuming as above that they have been encapsulated into objects). The action consists of beginning at the hypothesis, determining that it is true, and then "crossing the bridge" to the conclusion and asserting its truth.

Finally, there is a coordination of the function schema, as it applies to an implication valued function Q (obtained from a proposition valued function P) and the logic schema as it applies to the process modus ponens. Included in the function schema is the process of evaluation, that is, sampling values n of the domain (positive integers in this case) and computing the value of the function, $Q(n)$, that is, $P(n) \Rightarrow P(n+1)$. Suppose that it has been established that Q has the constant value true. The first step in this new process which must be constructed is to evaluate P at 1 and to determine that $P(1)$ is true (or, more generally, to find a value n_o such that $P(n_o)$ is true). Next, the function Q is evaluated at 1 to obtain $P(1) \Rightarrow P(2)$. Applying modus ponens and the fact (just established) that $P(1)$ is true yields the assertion $P(2)$. The evaluation process is again applied to Q but this time with $n = 2$ to obtain $P(2) \Rightarrow P(3)$. Modus ponens again gives the assertion $P(3)$. This is repeated ad infinitum, alternating the process of modus ponens and evaluation. Thus we have a rather complex coordination of two processes that leads to an *infinite* process.

This infinite process is encapsulated and added to the proof schema as a new object, proof by induction.

Predicate calculus. Quantification, which is the key mathematical process in the predicate calculus is one of the most important mathematical concepts. It is easy to make a long list of mathematics topics which appear to rely on it (at least mathematically and, as I am discovering, psychologically as well). These include the notion of limit, solving equations, linear independence, the difference between a proof and an example, etc. Indeed it is hard to imagine doing any of the mathematics that has arisen in the last 100 years without a strong notion of quantification.

On the other hand, this concept appears to be extremely difficult for students. Many teachers (even at the post-secondary level) are less than comfortable with it and it is rarely dealt with explicitly in the curriculum.

As an example of the difficulties that students have with quantification, consider the following two tasks which, together with B. Cornu, I recently presented to students at a Lycee in the terminal year. Using the method of written questions and answers followed by interviews of selected students, we asked the students to explain why each of the following statements was true or why it was false.

> For every positive real number a there is a positive real number b which is less than a.

> There is a positive real number x such that for every positive real number y, it is the case that x is less than y.

A fair number of students seemed to understand that the first statement is true, but very few realized that the second is false. Indeed many wrote (and stuck to it in the interview) that the second was true because it was the same statement as the first!

We then repeated the observation with a different group, this time reversing the order of the statements. This time many students thought that the first statement (which they saw second) was *false* because it was the same as the second (which they saw first).

One can conclude that, typically, these students tried to find the "easy way out" in considering the statement which they saw second. The fact that this led them to an error that they could not correct in the interview, together with the large number of errors made altogether, suggested to us that they did not have a very strong understanding of what I call (see below) a two-level quantification. It is hard for me to see how these students can get much out of the formal discussions of the definition of limit (which is a major part of their curriculum), given that this definition involves a three-level quantification.

In the genetic decomposition of the predicate calculus that I have derived, the schema comes from a reconstruction of the schema obtained by coordinating the schema for first order propositional calculus with the function schema. The construction is illustrated in Figure 3. The objects in the propositional calculus schema are the propositions. The most important process is the determination of the truth or falsity of a proposition. Other processes include the formation of new propositions by the standard logical operations such as conjunction, disjunction, implication and negation. They also include the process of expressing a natural language statement in the formal language of symbolic logic and translating from that syntax back to English. Then of course there are all the usual tasks that students are asked to perform such as manipulation of the formulas, construction of truth tables, determination of the validity of arguments and so on. Finally, we can mention the process of reasoning about a statement; for example, to know if the truth or falsity of the statement

$$(P \Rightarrow Q) \vee (Q \wedge R)$$

is determined once you know that $P \Rightarrow R$ is false.

Amongst the various manipulations of logical expressions, one in particular is important. That is the process of applying the conjunction operation ("and" or $\wedge$) to a set of propositions as in

$$(x_1 > b_1) \wedge (x_2 > b_2) \wedge \dots \wedge (x_n > b_n).$$

There is a similar process for disjunction ("or" or $\vee$).

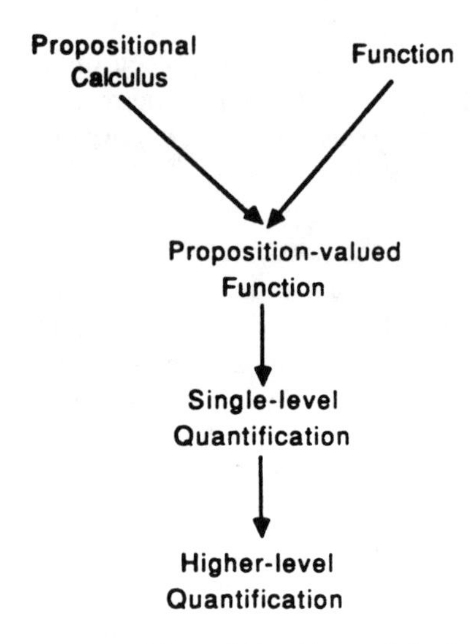

Figure 3. Genetic Decomposition of Predicate Calculus

In a sense, the objects in the first order propositional calculus are constants. In an expression such as $(P \Rightarrow Q) \vee (not(Q \wedge R))$ the quantities P, Q and R are constants whose values may be unknown, but fixed. The subject's thinking about such matters is elevated to a higher plane when the propositional calculus schema is coordinated with the function schema (appropriately reconstructed on this higher plane) to consider such an expression as determining a function--in this case of the three variables, P, Q and R. This is the beginning of the predicate calculus schema. Of course, a part of this coordination and reconstruction was discussed already in the previous section for the special case of proposition valued function of the positive integers.

As before, an important new process that can be constructed is the iteration (in the subject's mind) through the domain of a proposition valued function checking the truth or falsity of the resulting proposition for each value of the variable. Consider, for example a statement such as

> Given a car in the parking lot, if the tire fits the car, then the car is red.

Here, *tire* may be considered to be a constant, but *car* should be thought of as a variable whose domain is the set of cars in the parking lot. There is an obvious action of walking through the parking lot, checking each car to see if the tire fits and, if it does, seeing if the car is red. When such a statement appears in a mathematical context, as in

Given $x \in$ domain (F), if $|x - x_o| \leq \delta$, then $|F(x) - F(x_o)| \leq \in$

then the mental process consists of looking at each $x \in$ domain (F), checking that $|x - x_o| \leq \delta$ and, if so, seeing if $|F(x) - F(x_o)| \leq \in$.

This iteration process may now be coordinated with one of the two processes we mentioned earlier: applying conjunction or disjunction to a set of propositions. The resulting process can be encapsulated leading to a single existential or universal quantification as in

> There is a car in the parking lot such that if the tire fits the car, then the car is red.

or

$$\exists\, x \in \text{domain}(F),\ |x - x_o| \leq \delta \Rightarrow |F(x) - F(x_o)| \leq \in$$

We call this a *single-level quantification*.

The single-level quantification creates new objects which are again propositions so that all of the previous processes of logical operations, negation and reasoning about statements are reconstructed on this higher plane. Particularly important for understanding mathematiccs is the interiorization of a statement given as a quantification. It seems that the subject needs to have a strong mental image of the iteration and quantification processes that we have described in order to relate the statement to the mathematical situation that is being considered.

Next comes *two-level quantifications* in which two (usually different type) quantifiers are applied in succession to a proposition valued function of two variables. For example, the statements we have considered may be extended to obtain,

> For each tire in the library, there is a car in the parking lot such that if the tire fits the car, then the car is red.

or

$$\forall \delta > 0\, \exists\, x \in \text{domain } (F),\ |x - x_o| \leq \delta \Rightarrow |F(x) - F(x_o)| \leq \in$$

The process which we just described for constructing single-level quantifications ended with an encapsulation so that the result becomes a proposition which is a mental object. Note that the effect of a quantification is to eliminate a variable. If the original proposition valued function had two variables, then the resulting object actually depends on the value of the other variable and the schema for single-level quantifications can again be applied to this proposition valued function. For example, in the case of the tires and cars, the existential quantification over cars results in a proposition valued function of the single variable, *tire*. This function is then subjected to a universal quantification to obtain a single, constant proposition. Thus, when analyzing a statement which requires a two-level quantification over two variables, the subject begins by parsing it into two quantifications. There is an inner quantification over one of the variables in a proposition valued function of two variables. There is also an outer quantification over the other variable. What we have described is a coordination of these two quantifications to obtain a proposition which will be a two-level quantification. One way in which to proceed to higher-level quantifications is to again encapsulate this process to obtain a new object. Once it is encapsulated, it can then be subjected to the same process (thereby generalized) as were the single level quantifications.

Given a statement which is a three-level quantification, such as the definition of continuity of F at x_0,

$$\forall \in > 0, \exists \delta > 0 \ni \forall x \in \text{domain}(F), |x - x_0| \leq \delta \Rightarrow |F(x) - F(x_0)| \leq \in ,$$

the subject can group the two inner quantifications and apply the two-level schema to again obtain a proposition which depends on the outermost variable ($\in$ in this case). This proposition valued function is then quantified as before to obtain a single proposition. The entire procedure can now be repeated indefinitely to obtain quantifications of any level. At each level, the same processes of logical operations, negation, reasoning, etc. are reconstructed.

Functions. As I have said, my investigations of the construction of the concept of function are only beginning and I do not yet have a fully developed genetic decomposition for this concept. The ideas that are presented in this section are preliminary thoughts on which I am trying to base a first analysis of the concept. This is my attempt to develop the adult model referred to in the section, *Analysis of the concept*.

For most students, and indeed for many scientific workers, the idea of function is completely contained in the "formula." If you ask students for an example of a function, you will often get an algebraic expression such as $x^2 + 3$ with no mention of any kind of transformation. Just as with the concept of variable in which the student insists that x "stands for" a *single* number (which may not be known), the concept of function as formula has a very static flavor.

There are a number of ways in which such a function schema is inadequate. For one thing, the objects are restricted to those functions which can be conveniently expressed with a formula. This may suffice for elementary mathematics but it will not do for advanced mathematical thinking. When a function is the same as a formula, the action of evaluation on this object consists of plugging in numbers for letters and composition of two functions is restricted to substitution of a formula for each occurence of a letter. The notions of domains and range have no place in this schema and graphs, while manageable in themselves (because of their concrete and visual nature), have no connection with functions for the student with a function-as-formula schema. When the graph cannot be "seen" (as is the case with the characteristic function of the irrationals), then the student is unable to think about it.

A more powerful schema for functions will involve interiorization of actions. When a subject perceives a function, then he or she may view it as an action on objects that transforms them into other objects. This action can be interiorized. Thus I suggest that, to know a function is to construct a process corresponding to a situation which gives rise to it. Some may refer to this as a mental representation of the function, but we prefer to avoid such terminology because of its tendency towards the misleading suggestion that the internal process is a copy of some "external reality." The important point is that when a function is known as an interiorized process, then this knowledge has a dynamic flavor which affects the nature of the subject's manipulations of the function. Thus if a subject responds to a problem situation by constructing an internal process and uses this process to deal with the situation, he or she has a tool which is more powerful than the earlier, less sophisticated function schema.

Evaluation can now become the action of taking a particular value (in the domain of the function) and performing the process on it to obtain a new value (in the range of the function). It is then possible for the subject to coordinate a function's process and its graph. That is, there is the understanding that the height of the graph of a function f at a point x on the horizontal axis is precisely the value $f(x)$. The subject can then relate to the full power of graphing which is the relationship between the (static) physical shape of the graph and the (dynamic) behavior of the function.

Several important ideas in mathematics correspond directly to doing some of the things we have discussed with the process of a function. For example, the coordination of two processes corresponds to the composition of the functions. A function's process can be reversed, thereby obtaining the inverse function. It is by reflecting on the totality of a function's process that one makes sense of the notion of a function being *onto*. Reflection on the function's process and the reversal of that process are necessary for the idea of a function being *one-to-one*.

Another example of difficulties with reversing processes arises in the concept of composition of functions and solving certain kinds of functional equations. On several occasions I have given subjects the following kinds of problems relative to three specific functions, F, G, H.

1. Given F, G find H such that $H = F \, o \, G$.
2. Given G, H find F such that $H = F \, o \, G$.
3. Given F, H find G such that $H = F \, o \, G$.

Of course the first is much easier than the other two, and I find invariably that the third is harder than the second. I can suggest an explanation derived from the theory. The first kind of problem requires only the coordination of two processes that, presumably, have been interiorized by the subject. The second, however requires that the following be done for each x in the domain of H.

2a. Determine what H does to x obtaining $H(x)$.

2b. Determine what G does to x obtaining $G(x)$.

2c. Construct a process that will transform $G(x)$ to $H(x)$.

The third kind of problem can be solved by doing the following for each x in the domain of H.

3a. Determine what H does to x obtaining $H(x)$.

3b. Determine value(s) y having the property that the process of F will transform y to $H(x)$.

3c. Construct a process that will transform x to such a y.

Comparing 2b. with 3b. (the only point of significant difference), it appears that 2b. is a direct application of the process of G whereas 3b. requires a reversal of the process of F.

It is perhaps interesting to note that the difference in difficulty (between 2 and 3), which is observed empirically and explained epistemologically, is completely absent from a purely *mathematical* analysis of the two problems. They are, from a mathematical point of view, the calculation of $H \circ G^{-1}$ and $F^{-1} \circ H$, respectively, which appear to be problems of identical difficulty. This seems to be another important example in which the psychological and mathematical natures of a problem are not the same.

Another situation in which relative difficulty can be explained by the requirement of reversing a process occurs in the development of children's ability in arithmetic. According to Riley, Greeno and Heller (1983, p. 157), "Problems represented by sentences where the unknown is either the first (? + a = b) or second (a + ? = c) number are more difficult than problems represented by equations where the result is unknown (a + b = ?)." Obviously the first two problem types involve a reversal of the process which, in the third type, can be applied directly.

A number of important mathematical activities seem to require that the function schema be reconstructed at yet a higher level where a function is not only an interiorized process, but as a result of encapsulation, this process can be treated as an object by the subject. One way of representing functions mathematically that can help is the set of ordered pairs (with the "uniqueness to the right" condition) and another is the graph. In order for a function to be the result of a mathematical activity (such as solving a differential equation or setting up an indefinite integral) it must be an object. Similarly, it seems to me that the elements of a set must be (cognitive) objects and so that all of functional analysis with its sets and even structured spaces of functions depends on the object nature of a function.

At the same time, and this may be a further reconstruction of the function schema, it seems that in many situations the subject needs to think of a function simultaneously (or at least in rapid succession) as both a process and an object. Consider, for example, the various binary operations of functions such as pointwise addition, pointwise multiplication or composition. In reflecting on the addition of two functions, the subject must see this as a binary operation which takes two objects and transforms them into a new, third object. To actually do this, however, the original two objects must be "decapsulated" back into processess, these two processes coordinated (by means of "pointwise addition") and the resulting process encapsulated into an object which is the new function that appears as the result of the operation of addition.

As a final example, consider how complex, in these terms, is the following statement which is straight-forward from a mathematical point of view.

> In the semigroup hom(G, *o*) of endomorphisms of a group G under the operation of composition, the subset of those endomorphisms which are isomorphisms form a group.

To understand this statement (and check that it is true) I think that the subject must think of functions as objects since they form a set, and later a subset, and then understand composition as I have described it to get a firm grasp on hom(G, *o*). Now, in dealing with the group axioms, the cognitive interpretation of function needs to go back and forth between process and object. The two interpretations must be coordinated in order for the subject to grasp the somewhat subtle idea that the group identity is the identity function and the group inverse of a function is its function theoretic inverse--and to understand that this connection is not exactly an accident.

Other Mathematics Topics

As a final set of examples, I will mention a few isolated topics in mathematics and show how one might think of them in terms of the constructive aspects of reflective abstraction that I have described. The material in this section is purely speculative and has no relation to any observations that I have made within the research paradigm put forth. I have two purposes in making these speculations. The first is to point out that the kinds of constructions I have presented can be used to describe a wide variety of mathematical concepts. The second is to suggest that it will be possible to at least begin the application of my research program to every mathematics concept in the undergraduate curriculum, by using the theory that I am developing to make an initial analysis of the topic.

Interiorization can occur in connection with the notion of linear independence of a set of vectors. In order to understand the definition of this concept, the subject can interiorize the act of forming finite linear combinations of the vectors and checking the result, imagine repeating this act exhaustively, and determine that, with one exception, the result is never zero. Once this process is present in the subject's mind, it is possible that he or she could imagine attaining zero with non-trivial coefficients, resulting in linear dependence.

A completely different example of interiorization arises in understanding proofs. When the mathematician exclaims (as which of us has not?) that "I can understand each step of the proof, but I don't see the whole picture," he or she is expressing the necessity of interiorizing a whole collection of processes and coordinating them to obtain a single process. The interiorization of the total process is, in my opinion, the final step in "making a proof your own."

Interiorization is not always difficult. Most students have little trouble with constructing a mental process for multiplying a matrice and a vector, or two matrices. I think that this is because there is a straightforward "hand-waving" action, used by most teachers, that is a physical representation of the multiplication and forms an intermediary between the external process and its interiorization. Mathematics becomes difficult for students when it concerns topics for which there do not exist simple physical or visual representations. One way in which the use of computers can be helpful is to provide external representations for many important mathematical objects and processes.

A whole class of examples of coordination of schemas in advanced mathematics is given by the "mixed" structures: topological vector spaces, differentiable manifolds, homotopy groups, etc.

Returning to the realm of high school mathematics, one can often see a fairly shallow understanding of the concept of variable as little more than a letter (as opposed to a number) which can be manipulated according to certain rules that must be memorized, and may stand for a single (possibly unknown) number. From the point of view of my theory, the idea of variable is an encapsulation (and this is where the object comes from) of the action of substituting (infinitely) many numerical values for the variable (and this is where the manipulations come from--ordinary arithmetic).

The indefinite integral forms an important example of encapsulation together with interiorization. Estimating the area under a curve with sums and passing to a limit is, of course, a process. Students who seem to understand this often have difficulty with the next step of varying, say, the upper limit of the integral to obtain a function. What is lacking is the encapsulation of the entire area process into an object which can then vary as one of its parameters vary. This then forms a "higher-level" process which specifies the function given by the indefinite integral. The complexity of this total process perhaps helps explain why students have such difficulty with not only the Fundamental Theorem of Calculus, but such powerful definitions as

$$\log x = \int_1^x \frac{dt}{t}, \quad x > 0$$

A rather pervasive example of encapsulation in mathematics is duality. The dual of a vector space, for example, is obtained by considering all of the linear transformations from the space to its scalar field as objects, collecting them in a set, and introducing an algebraic structure on this set.

The simplest and most familiar form of reflective abstraction is generalization. The schema of factorization of positive integers can be generalized to factoring polynomials, and then to an arbitrary Euclidean ring. Vectors of dimension two and three are generalized to include higher, and even infinite, dimensional vectors.

Finally, I can mention a number of familiar activities in mathematics that involve the reversal of a process: subtraction and division, solving an equation, logarithms, proving an inequality (in which one often starts with the conclusion, manipulates until something known to be true is obtained, and then sees if the argument can be reversed), and the mysterious choice of expressions such as $\frac{\epsilon}{2\sqrt{M}}$ in proving limit theorems.

Notes

I would like to thank Les Steffe for an extensive interaction that has been very helpful to me in developing my ideas and in setting them down on paper.

10 Reflective Abstraction in Humanities Education: Thematic Images and Personal Schemas

Philip Lewin

Introduction

Paul Ricoeur (1967) has observed that "The symbol gives rise to thought" (p. 348). In this way, he points to the hermeneutic dimension of symbols. Symbols confront us as givens, as already-there representations that coalesce prior, and perhaps non-conscious, reflection in a concrete image. In so confronting us, they lead our thinking, enticing us to further reflection.

My experience in the classroom illustrates Ricoeur's claim. At the end of each semester, I invite students who have just completed a required humanities course to discuss a book with me that provoked or stimulated their thinking (not necessarily one that they liked). Not invariably, but often enough, an interview with the following features occurs: A student will like or dislike a book for this or that reason - he may like how Socrates, in *Euthyphro* and *The Apology*, makes people uncomfortable, exposing their purported wisdom as shallow opinion, or she may not like *The Prince*, seeing Machiavelli's portrayal of humans as too cynical. Later, after the conversation has moved to more general concerns, the student will introduce an observation thematically relevant to his or her earlier points about the book - that he is a confirmed Catholic with many questions about his faith who has always been uncomfortable with people who seemed unduly confident in their religious convictions, or that she is tired of how boyfriends manipulate their girlfriends, playing by a double standard.

These interviews have several interesting dimensions. First of all, there is their unexpectedness. I cannot predict which texts will appeal to a particular student, and which theme in the text will occasion a response. I, for instance, teach Socrates as a gadfly in Athens, as ambiguously content to be condemned to death, as a figure who forces us to consider the limitations of critical discourse in a democracy; I underplay his baiting, provocative style. And it has never occurred to me to think that there is a relation between Machiavelli and adolescent love. But my own presuppositions notwithstanding, these are the kinds of themes of which students spontaneously speak. And once they are brought to my attention, I can acknowledge them. They are not fabricated, but legitimately reside in the text, waiting like so much enchanted treasure for the light touch through which an inquiring mind can set them free.

Why would students choose such apparently simple themes as these to discuss? The answer is not one of cognitive deficiency, that they just aren't smart enough to see the more sophisticated issues. On essay examinations and reflective, critical essays, when asked to perform to a teacher-set standard, they can measure up. They can discuss Socrates' inner voice and the betrayals of Alcibiades; they point out the deep Christian conviction behind Machiavelli's apparent hypocrisy. The answer is not cognitive deficiency, it is not laziness. It may simply be that, after an interval of weeks or months, these themes are what stuck. Inquiring minds are not empty. They are more like velcro than *tabula rasa*, possessed of "hooks" on which an image from a text may get snared as it is shredded through. Maybe only the image gets caught, torn asunder from the text. Maybe a chapter dangles on, or the whole book, connected, at least at first, only by that image. Maybe that's enough.

I believe an argument can be made that students' appropriation of a text may take place in counterpoint to epistemically salient thematic images provided by the text. Texts provide themes and images that guide students in an imaginal series of cognitive acts, which I will discuss under the Piagetian rubric of 'reflective abstractions.' Conversely, the images made available through the text are mediated by students' prior personal concerns which are organized in what we can call "personal schemas." T.S. Eliot (1920), in discussing the problem of adequate motivation in Shakespeare's *Hamlet*, introduced the concept of the "objective correlative" as "a set of objects, a situation, a chain of events which shall be the formula of that particular emotion; such that when the external facts, which must terminate in sensory experience, are given, the emotion is immediately evoked" (p. 100). (Eliot said Shakespeare had failed to provide Hamlet with an objective correlative for his melancholy and procrastination.) Leaving aside Eliot's argument, I would appropriate his concept of objective correlative for my own purposes, slightly modifying it so it connotes those circumstances external to a knower whose significance could be construed by the knower as sufficiently important to induce a particular affective response. I would suggest that responses by students to texts indicate that texts function as objective correlatives. They provide some motif of personal relevance through which the student can gain access to the text as the text assumes a kind of objective importance to the student. The symbol in this way gives rise to thought.

In this essay, I would like to explore this insight, particularly as I think it applies to the experience of students in classrooms. I assume that my students, at least in their freshman year, are novice knowers who may lack not only sophisticated understandings, but the critical skills through whose application sophisticated understanding becomes possible. As an educator, my central concern is to promote those skills. My means of doing so is through reflective encounters with texts.

The first part of my presentation will be theoretical. I will begin by suggesting a reading of the constructivism of Jean Piaget that emphasizes how the epistemic object is situated with respect to an epistemic subject. I will then try to place Piaget's concerns more centrally into the context of education and indicate how education may be conceived hermeneutically. Part of my concern in the first part is to indicate areas of overlap among education, epistemology, and contemporary studies of narrative whose basis is their common assumption of constructivism.

The second part of the paper will offer illustrations of the perspectives introduced in the first part through examination of student protocols. I will for the most part restrict my scope from the larger guiding images of which I have just spoken to smaller, more manageable ones, gaining in precision what is lost in scale (though I will conclude with one extended example). I will use Piaget's theory concerning the process of reflective abstraction to model what might be happening in the mind of a knower as she (or he) constructs her (or his) understanding of a text or artifact.

Piaget, Education, and Hermeneutics

Piaget distinguished between what he called the figurative and the operative aspects of thought in order to differentiate that aspect of cognition concerned with the static representation of an epistemic object from that aspect concerned with transformations between states. The figurative derives initially from the progressive imitation of external objects through sensori-motor and perceptual activity (accommodation), an internalization that is at first uncoordinated and not epistemically self-conscious. Though the figurative is constructed over time, the epistemic subject at first is unaware of this developmental history. The reconstruction of activity on the plane of thought through interiorization results in operational knowing, within which the figurative is grasped as a sedimentation of epistemic activity. It is comprehended as both the outcome of earlier transformations and a presently stable configuration which subsequent operations will transform. The figurative is not the source of knowledge, as it would be if perception were a simple, passive imitation of the external world. Rather, it serves a limit function, demarcating the boundary at which the free play of assimilation finds resistance. The figurative thus both confronts thought as a given and guides further operational transformations much as, following Ricoeur, symbols do.

Piaget's exploration of the dialectic between the figurative and the operative was deeply intertwined with his concern to elucidate the grounds of certainty of logico-mathematical knowing. He tended to pursue the investigation of those figurations which were empirically derived as opposed to those which originated in internal imaginal processes. He assumed that only empirically derived figurations - perceptions, the mental image of deferred imitation, the reproductive image of memory - could provide the link between logico-mathematical knowledge and its application to the external world. As he observed in *The Mechanisms of Perception*, "The essence of the operational concept of intelligence is to negate the existence of any radical dualism between experience and deduction....Logico-mathematical knowledge does not detach us from reality or from the world of objects, but only enlarges that world by incorporating it into the set of all possible events" (p. 358).

However, there is a rich array of imaginative figurations, more properly hermeneutic, which although grounded in the empirical and contingent, nonetheless resist the perfect regulation of logico-mathematical deduction. Piaget laid the basis for considering such figurations in his book on symbols, *La Formation du Symbole* (translated as *Play, Dreams and Imitation in Childhood*). There he introduced the concept of the "affective schema" which, like Freud's defense mechanisms, provide "relatively stable modes of feeling or reacting" (p. 206) by which "personal schema" of the knower can be constructed. Though Piaget did not further develop his insight into the affective schema, I believe it represents an extremely important concept, for it connects the constructivism of a generalized epistemic subject to the particular constructions which actual knowers make. The affective schema arises in the assimilation of aliments (that is, that which is epistemically novel) whose accommodation comes to take on personal importance for the knower.

Central to Piaget's thought is the recognition of the affect as motivator and energizer for the cognitive. Within his preferred domains of logico-mathematical construction, the affective manifests itself in the experience of felt necessity which accompanies operational intelligence. Piaget pointed to the feeling of necessity as evidence that reversibility of thought, signaling a new competence and not merely serendippitous performance, had been achieved. Building on this idea, but transferring it from domains of the logico-mathematical to domains of the imaginal, the affective schema can be seen as engaging truth not as a universal condition for all epistemic subjects, but as personal and particular to concrete individuals.

What makes the affective schema of interest from both an epistemological and a pedagogic point of view is that the process of its construction is also the process in which personal concerns of the knower may come to a focus of clarity and be recognized as such for the first time. Prior to its construction, these concerns may exist in inchoate form, as fragments or tendencies not yet organized in a coherent set of concepts or symbols. The process in which they achieve a cognitive coherence can thereby be a process of personal integration for the knower, an integration which is a central moment of education.

Many of us who are involved with education in the humanities and social sciences understand that such education is fundamentally a *paideia*, an education whose central concern is with the formation of character. The chief means at the disposal of humanities educators to foster this development are the various resources through which other individuals have expressed their own *paideia* - literary, musical, visual, analytic, performative, and so forth. We have access, that is, not to these individuals in the moments in which they create, but to their artifice, not to a life lived in its ongoing fullness, but to a stable presence, the artifact, generated out of that ongoingness.

We ask students to engage these artifacts. It is our belief that through such encounter, they will be challenged, enriched, educated in the literal sense of finding themselves drawn out. At the extreme, such education produces not only the private self, but the public character, the knower as citizen, the Aristotelian ideal of the fully empowered participant in the culture. This ideal underlies the progressive pedgagogy of Dewey (1916), and has more recently informed such critiques as that of MacIntyre (1981) and Bellah, et al. (1985). It is irreducibly dialectical; the sense of self emerges out of encounter with the artifact. Understanding self and understanding world become differentiated moments of the same acts of construction. Each ineluctably requires the other.

The dialectical model of the educational process stands in marked opposition to those models which assume either an autonomous knower or an autonomous artifact, or both. If we posit an autonomous knower, then education becomes accoutrement, an embellishment of cultural literacy taken on by an imperious Cartesian *cogito*. In such a model the self-interested individual engages the text as it engages the other, freely, as an independent agent, accepting or rejecting the artifact as it is useful or not for the self's own projects, sublimely independent of facilitating conditions. The artifact is one more commodity in the marketplace of ideas; no one is forced to buy. Education is patina, icing, the couthing of the uncouth, in which individuals partake or not, as they choose.

If, on the other hand, we assume an autonomous artifact, we find that the artifact functions like an arbitrary monarch, demanding that the knower conform to it the way we might assume a tyranny, independent of the general will, carries on largely through its own momentum. The role of the knower, like that of the subject, is ambiguous at best and negligible at worst. Knowers are to be co-opted, willingly or not, with greater or lesser degrees of awareness, with greater or lesser degrees of corresponding alienation and cynicism. The knower is a passive container, to be processed and neutralized.

Education is forced conformity, the stuffing of the free spirit with the ideological residue of the powerful and the corrupt.

Both of these alternatives are unsatisfactory. In both, knower and artifact have no essential need of the other; all relations between them become relations of power and self-interest. In contrast, education as *paideia* offers the means for a conversation between self and text to take place, a subtle and mutual negotiation whose outcome can be the responsible co-collaboration in which texts are interpreted and lives are constructed. And, as we will see below, such negotiation involves others as well. This triadic process of social construction - involving self, other, and text - generates the interpretive communities which simultaneously impart meaning to texts, co-construct culture, and provide components of identity for their members.

Insofar as education as *paideia* explores the interface between artifact and the construction of self, it has important connections with current thinking about the nature of narrative, particularly with the recognition of the role narrative plays as people construct their understanding of their own lives (see, e.g., Bruner, 1987; Cohler, 1982; Csikszentmihalyi & Beattie, 1979; Freeman, 1984). The constructing of narrative could be a fundamental mode of how we are human, what Heidegger (1962) called an *existentiale* of *Dasein*. In a Kantian sense, our status as mythopoeic beings, as story shapers, might be the transcendental condition for the possibility of all experience. (Though I cannot pursue this point here, narrative as a mode of understanding offers a dynamic synthesizing of both what Kant conceptualized as static and separate categories, and what Heidegger identified as the three *existentiale* of understanding, mood, and speech.) There is no necessity that the narratives we construct over the course of a life be mutually consistent, only that we engage in the narrative act.

We come to comprehend the world and our place in it by embedding the events of everyday life into larger structures, emplotments, of coherent experience. These larger narratives sustain our immediate sense of who we are; at the same time, these narratives are reconstructed throughout life, not only incorporating new experience, but reconceptualizing old. The early narrative of academic success is undermined by subsequent events of failure; but in time, the narrative of defeat again becomes one of success as early failure is re-understood as having provided the proving ground which sustained subsequent victories. Our present is constructed on the basis of having conceived a significant past; our past is reconstructed on the basis of the perquisites of the present, in a dialectic that lasts a life, that encompasses many stories.

At the same time, the stories we tell are not fantasies composed without regard for the possibility of corroboration from others. The construction of self-narratives is mediated through our social interactions. To use Coleridge's distinction, self-narratives are works of imagination, not of fantasy. They are composed responsibly, not whimsically, in response to external circumstance (including the constraints of audience expectation), not despite it. Our relation to external circumstance functions in the personal tales we tell much as do Winnicott's (1971) transitional objects. Transitional objects are neither totally objectified, detached from the knower, nor are they totally subjectified, extensions of infantile omnipotence. Winnicott suggested that these objects take on a privileged character for the infant. They serve as ambiguous markers around which the boundaries of 'the me' and 'the not-me' can be negotiated, through which the child can gradually come to know itself-in-the-world. Our personal narratives function in a similar way. They recognize the independence of the other and the objectification of the world - the constraints and demands they impose - while simultaneously imbuing other and world with significance in terms of personal experience, embedding them within our plots.

The importance of this doubleness can be seen if we consider Heidegger's (1962) description of our modes of engagement with the worldhood of the world. Our naive encounters with the world are in the mode of the ready-to-hand, the mode within which the world exists as an extension of our projects. Using Heidegger's example, we observe that tools exist as extensions of our bodies, not noticed in themselves, but understood as they are incorporated to the performance of a task. When something goes awry - when the tool breaks, when the obtrusiveness, the obstinacy of the world announces itself - the mode of the unreadiness-to-hand forces our attention. We then move into the mode of the present-at-hand, in which "things," formerly assumed as transparent, take on the character of "objects," as that-which-stands-against us. We notice the object, become aware of it in its own right, and in that moment of disequilibration may come to understand aspects of it of which we formerly were not conscious. If we then return to the ready-to-hand and re-incorporate the object as a tool to a further end, we do so with a greater knowledge that may facilitate our re-engaged performance.

The present-at-hand is the typical mode of scientific objectification, the mode in which attempted characterization of the thing-in-itself, as it "really" is, takes place. The present-at-hand is a mode of disengaged knowledge, of knowing removed from context - from contexts of use, of power, of bias, of personal preference, of gender, of history. In our culture, it is the preferred mode of regarding knowledge. Notice, though, that it is a secondary mode, derived after an original engagement with the world. In terms of our personal projects, the present-at-hand is like an enormous wave, momentarily dominating our horizon until it is submerged again in the deep and steadfast ocean of the ready-to-hand, of our lived-engagement with the world. I take account of my broken tool only long enough to fix or replace it, until I can again immerse myself in my initial project. Even in those moments when the ready-to-hand has become a present-at-hand, a focus of attention, the present-at-hand is itself understood through a prior configuration of a ready-to-hand. The sensations aroused as I run my fingers over the roughened edge of the hammer head are not of interest in themselves, but are taken as indications of the serviceability of the hammer for my project. But the hammer, now present-at-hand, is itself understood only through the running of my fingers, ready-to-hand. There is a necessary dialectic between the modes, between the present-at-hand focus of our projects (from lifelong ones like career and family to immediate ones like tying my shoes) and the ready-to-hand means through which they are achieved (education, courtship, interpersonal associations, having shoelaces), an infinite regress whose foundation can never be reached. But to speak of foundations and infinite regress is misleading; it suggests that our inability to reach a foundational present-at-hand, a rock-bottom fundament of reality that could be known and fully incorporated into the ready-to-hand, is somehow a tragic limitation on the human condition. Certainly the attempt to reach such a foundation has been replayed repeatedly in western thought, from Plato to Descartes to Husserl. And certainly the Derridean concern with *differance* (1982) is precisely the simultaneous recognition and unmasking of this quest, the paradoxical revealing of the inevitable absence of presence, of foundations that are endlessly deferred.

My point here, though, is that from the perspective of narratizing, of the embeddedness of the ready-to-hand in our lived projects, a position that prioritizes the present-at-hand is misguided. What is important is not the certainty of foundations, but the project itself. Or better, what is important is the certainty of the present-at-hand only insofar as an unreadiness-to-hand requires attention, only insofar as we seek to reincorporate the present-at-hand, objectified, back into the ready-to-hand. The movement in life is first from contextualization to objectification; but then necessarily from objectification to recontextualization.

If this model is accurate, it makes it all the more ironic that the mode of the present-at-hand is the mode in which the information we disseminate in classrooms is usually packaged. Far too frequently, pedagogy is undertaken in the mode that ignores, that covers over, precisely those contexts from which the present-at-hand originally derived that made it worthy of attention in the first place. The consequences of this covering-over become clear when the result of instruction is precisely the opposite of the intent; instead of achieving a *paideia*, students understand their classroom experience as some version of oppressive rote memorization whose meaning or importance is unclear or irrelevant. School is apperceived as a distraction to living. This is education where the autonomous, undigestible object confronts the autonomous, unwilling knower.

In contrast, hermeneutic theory emerging from Heidegger gives us a means of approaching this problem in a different way. Hans-Georg Gadamer (1975) speaks of texts as having a "horizon" of concern that includes the particular motivating factors involved in the author's creation of the text. A reader comes to the text, in turn, with his own horizon of "pre-judices," of questions and concerns. "The reconstruction of the question to which the text is presumed to be the answer takes place itself within a process of questioning through which we seek the answer to the question that the text asks us" (p. 337). The result is a "fusion of horizons" between the horizon of the text and that of the reader. Paul Ricoeur (1976) emphasizes this event-character of the fusion in what he calls an appropriation.

> Appropriation remains the concept for the actualization of the meaning as addressed to somebody. Potentially a text is addressed to anyone who can read. Actually it is addressed to me, *hic et nunc*. Interpretation is completed as appropriation when reading yields something like an event, an event of discourse, which is an event in the present moment. As appropriation, interpretation becomes an event. (p. 92)

Appropriation is the act that overcomes the distantiation of the text. It is the situating of a knower in relation to a text in such a way that the world of the text can be disclosed. In the possession of such worlds, the knower can expand and empower itself.

Let me link this discussion of narratizing to education by suggesting that in the classroom, artifacts and texts substitute for what in life are projects and lived events. They have the same epistemic function and elicit the same dynamic of response. What we as teachers ask of our students is that they engage texts as objects to live with and think with (as initially an unreadiness-to-hand), a resistance which must be approached consciously (as a present-to-hand). But we encourage this only to the further end of making texts "ready-to-hand." Texts, once assimilated by a knower, become internalized cognitive tools through which a lived relation of self to world can be maintained. The text, like Winnicott's "transitional object," serves the complex function of negotiating the boundaries between self and world. Simultaneously it affords access to the world with respect to the ongoing personal construction of meaning, a construction that is both socially motivated and socially embedded. Education, like living, is hermeneutic.

Reflective Abstraction of "Being Romantic"

Let me now try to apply some of these theoretical issues to teaching. I will discuss work done in connection with introducing the study of Romanticism in a required humanities course. I wish students to achieve at least a modestly sophisticated understanding of Romanticism, within which I would include seeing it as (1) a crystallization of certain ideas concerning human capacities (e.g., the essential self and the value of imagination) and the relation of humans to nature; (2) a dialectical response to Enlightenment rationalism; (3) a response to industrialization; and (4) a movement which, together with the Enlightenment, continues to structure how we conceive ourselves today.

Mine are not humanities students; few if any will do further work in the history of ideas, let alone read a poem of their own volition. They have little or no background in western culture, generally lack the cognitive skills requisite for close, analytic reading, and are usually deeply engaged in their various pre-professional curricula (engineering, science, management) in comparison to which a humanities course is a diversion at best, a despicable waste of time at worst. Yet I feel my goals are reasonable: to help students get a sense of Romanticism as a configuration of thought at a certain moment in western history, and to suggest to them how many of their assumptions (especially of the self conceived as sacred, unified, and essential, separate from constraints of class, gender, race, status, etc.) are a continuing legacy from that period. The key pedagogical step through which I try to achieve these goals is to ask students to juxtapose their pre-existing ideas about Romanticism with their growing classroom familiarity of it in order to facilitate the examination, and perhaps revision, of their pre-suppositions. The rationale for such a tactic is basic to a constructivist learning theory. Unless a student can find the relevance of a text, can construct its meaning for her or his own life, there is no possibility for *paideia*, for an engagement that enriches the self.

I would add that traditional learning taxonomies, such as Bloom's (1956), assume the already existing concerns of the student do not interfere with the acquisition of new material. That is, they assume that the absorption of new material takes place in a cognitive space isolated from the rest of the knower's knowledge and that integration of new material with old occurs only at "higher" levels. Recent research on cognitive representation, such as schema theory and work on problem-solving strategies by novices and experts (e.g., Anderson, 1977; Glaser, 1984; Norman, 1980), suggests this assumption is in error. Ironically, though, much of the theoretical perspective informing work on schemas is influenced by computational and information-processing models which continue to use an associationist epistemology. In contrast, I would argue that a constructivist epistemology is necessary if one is to properly take into account the already ongoing activity in the mind of the knower.

Piaget's theory offers us such an epistemology. At the center of Piaget's work is a fundamental cognitive process which he termed "equilibration" (Piaget, 1985). Speaking informally, we can describe equilibration as the process by which a knower attempts to understand something it notices (which we can refer to as a "cognitive aliment") by situating that aliment in its overall cognitive system. Such situating is successful as the knower cognitively constructs its understanding of the aliment through the process of "reflective abstraction." It is important to note that while equilibration takes place more or less automatically, reflective abstraction does not. That is, a knower may re-equilibrate by denying, implicitly or explicitly, that the aliment offers an occasion for re-thinking, for cognitive re-construction. Re-equilibration without reflective abstraction offers a means of understanding the common yet counter-intuitive case in which students declare a new competence or understanding that is belied by their performance, or alternatively, when students deny the relevance of an example or idea explicitly intended to challenge their current understanding.

Reflective abstraction is the process through which the cognitive activity of the knower is constructed and systematized. It entails two moments. First a cognitive structure must be reflected to a higher level of understanding from which it can be used more generally. Second, there must be some degree of mental reorganization, or reconstruction of the relevant parts of the knower's cognitive system, to appropriate this understanding within it. As the first moment promotes generalized application, so the second results in differentiated application, application that is appropriate in terms of the existing understandings of the knower.

In the task I wish to discuss, an introductory assignment on Romanticism, students were asked to perform three separate cognitive acts. (The assignment is reproduced in Appendix I.) First, they were asked to describe what it would mean for someone or something to be described as "romantic." This is a first moment in a potential reflective abstraction; students have been asked to encapsulate their understanding sufficiently to allow its use as a cognitive tool (using Heidegger's distinction, the ready-to-hand here becomes a present-to-hand). This first task, it should be noted, is naive; all it asks is a statement of the perhaps systematic, perhaps inchoate thoughts a student already holds about romantics.

Second, students were asked to read a short poem by William Blake (see "Mock on, mock on, Voltaire, Rousseau," also referred to as "The Scoffers," in Appendix II) and to paraphrase its meaning. They were told that this poem is generally considered a good example of Romantic poetry.

The cognitive process at work in reading a poem first entails internalization, or a mental imitation of the literal meanings of the words and sentences of the poem, and secondly, the reflective abstraction that Piaget called interiorization, by which a knower generates a coherent reading of the text from his or her internalization of it.

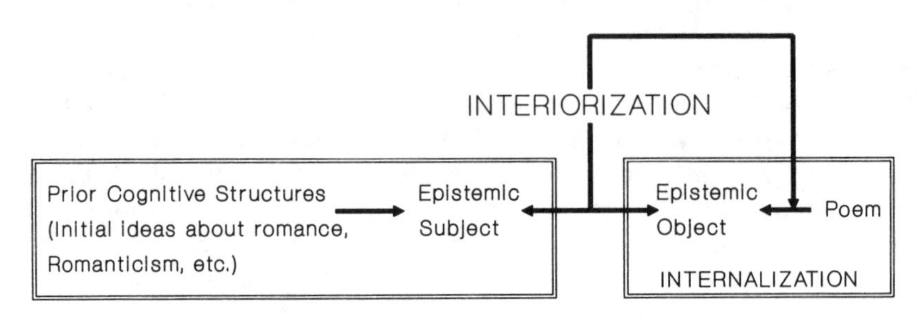

Figure 1. Model of initial epistemic condition.

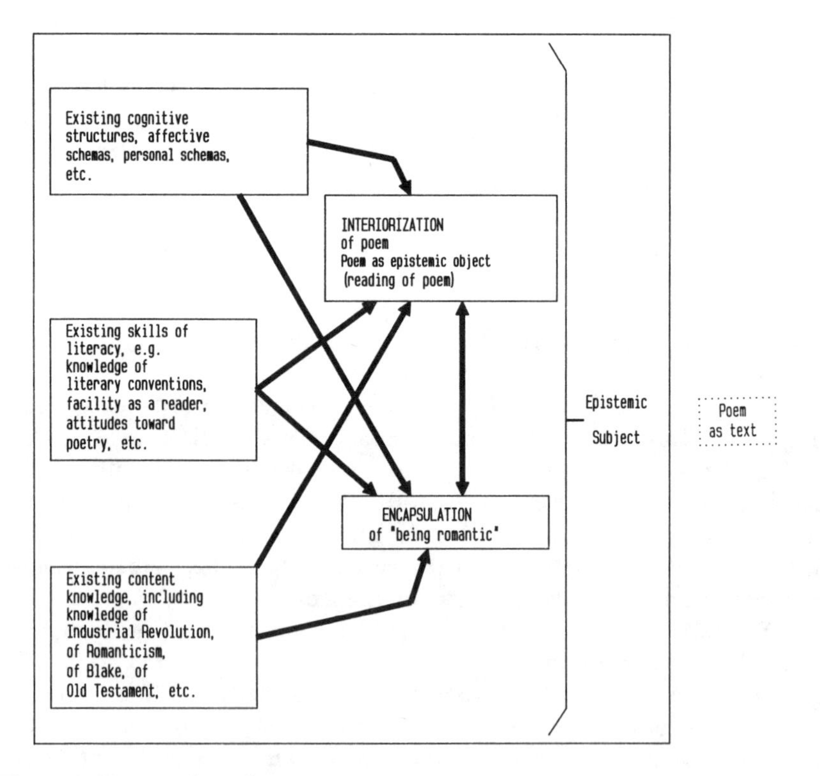

Figure 2. Interaction of structures.

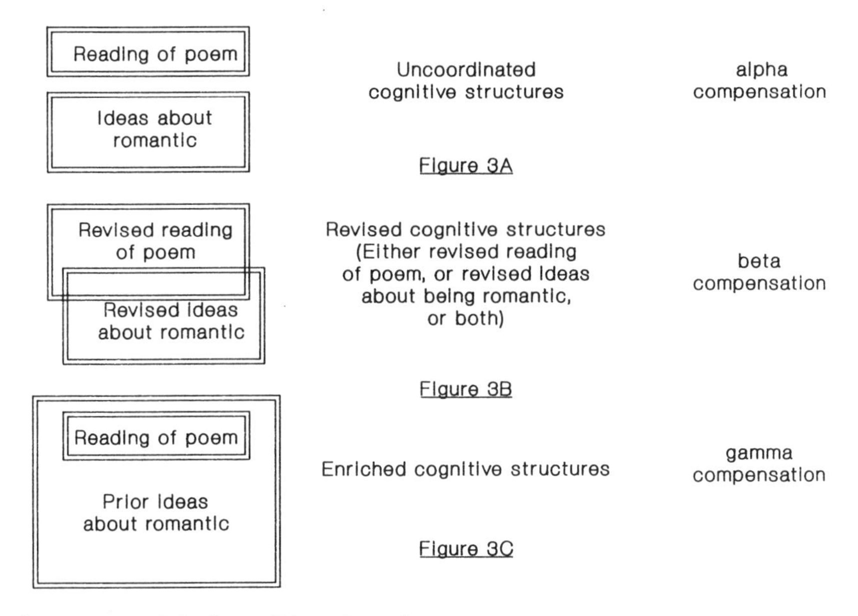

Figure 3. Model of possible epistemic outcomes.

I might mention that these processes are so poorly understood by traditional empiricism and associationist psychologies that they are presupposed as transparent. Each, however, is a complex cognitive act on its own terms, involving a number of individual reflective abstractions. In turn, internalization and interiorization are themselves dialectically related. That is, while the process of forming a literal, imitative mental representation of a cognitive aliment (the process of internalization) and the process of determining the non-literal signification of that representation (the process of interiorization) are epistemically distinguishable, they are experientially intermingled, each leading the other by turns in a co-construction of meaning. We could draw an analogy here to the neurophysiological process of perception. Like internalization, the transduction of energy differences detected in the ambient generates a seeing of shape, color, texture, etc. Yet this moment in perception is usually indistinguishable in experience from apperception, or the identification of such a manifold "as" something. This attribution of meaning corresponds to interiorization.

Wolfgang Iser (1974) suggests the kind of dialectic at play between internalization (or what he calls "the virtual dimension of the text") and interiorization (or "the imagination of the reader") in his description of the phenomenology of reading. He remarks that

> The literary text activates our own faculties, enabling us to recreate the world it presents. The product of this creative activity is what we might call the virtual dimension of the

> text, which endows it with its reality. This virtual dimension is not the text itself, nor is it the imagination of the reader: it is the coming together of text and imagination. (p. 278)

Below I will consider a few of the contributory factors operating in the construction of a virtual text. For the purposes of fulfilling this assignment, however, whatever internalization/interiorization of the poem a student achieved was acceptable.

Third, students were asked to try to apply their description of "romantic" to the poem, and see whether or not it applied. Cognitive activity consisted of the knower's generalizing his (or her) naive understanding of "romantic" in an attempted application. Blake's poem thus served abductively as both (1) object to interiorize and (2) object to which to apply the encapsulated concept of "romantic."

The encapsulation of "romantic" and the internalization and interiorization that jointly generate a reading of the poem are both instances of reflective abstraction. A third reflective abstraction, in which one or the other of the encapsulation or the reading were modified, may or may not occur. An illustrative case would arise if the application of "romantic" to the poem were imperfect; a disequilibration would then result and be acknowledged; and the disequilibration would promote a cognitive conflict whose outcome would be an enriched sense of either the poem, or of the meaning of "romantic," or both. That is, a possible reflective abstraction might consist of the encapsulation of "romantic" followed by a differentiation of the meaning of "romantic" to include both the original, naive meaning and the novel meanings aroused through the poem. Alternatively, it might consist of re-reading the poem in the light of one's sense of "romantic." Nothing would be lost or ignored; rather, apparent discrepancies between these cognitive objects would be resolved through differentiations or expansions of the sense of "romantic." At best, these resolutions would be accompanied by the affect of "felt necessity," of the sense of self-evident rightness which Piaget saw as the essential affective accompaniment of reflective abstraction.

Equilibration and the Cognitive Compensations

In *The Equilibration of Cognitive Structures*, Piaget (1985) differentiated three loci at which the disequilibration/equilibration process comes into play and three forms of cognitive compensations (alpha, beta, and gamma compensations) related to reflective abstraction. I will illustrate some of these concepts through examining student protocols related to their sense of "romantic." Figures 1, 2, and 3, if examined in sequence, show as a process the cognitive appropriations involved. However, let me begin by qualifying the limits of what this inquiry can demonstrate. First, novel aliments need not induce conscious cognitive activity in the mind of a knower, and cognitive activity need not be self-aware. Second, as investigators, we only occasionally have direct experience of an epistemic subject in the process of constructing his knowledge. More commonly, we infer on the basis of indirect evidence - such as changes in what people value or how they behave, what new projects they choose to undertake, or of positions they have revised, and so on - that some such process has taken place. As investigators, knowers once removed, we too are constructivists. Piagetian theory in its modeling of cognitive behavior is a hermeneutic.

The three types of cognitive compensations - alpha, beta, and gamma - chart equilibration through three degrees of increasing adequacy. Alpha compensation refers to those cases in which the knower effectively ignores potentially disequilibrating aliments. This results in reequilibration without reflective abstraction, with a potential proliferation of independent and unintegrated cognitive structures (see Figure 3A). Note that this state of ignoring disequilibration is very different from the ostensibly similar state of entertaining contradiction. In the latter case, two sub-systems are recognized as being in conflict, but occupy different loci within the cognitive system as a whole.

In beta compensation, perturbation by a cognitive aliment is acknowledged. Compensation occurs through accommodation; that is, the cognitive structures of the knower are modified or extended in a way that allows the disequilibrating aliment to be integrated without covering it over or ignoring it (see Figure 3B). Here reflective abstraction occurs; the assimilatory scheme is modified as it accommodates the object.

Finally, in gamma compensation, anticipations of possible perturbations already exist as part of the cognitive structure, so that potentially disturbing aliments are compensated in advance (see Figure 3C). As new aliments are assimilated, cognitive structures are enriched.

I should pause at this point to underscore the fact that the distinctions being made here are epistemic. They are abstracted from ongoing human projects. Within real contexts of human purpose, no epistemic compensation is inherently preferable to another. Thus, we can say that epistemically, gamma compensations are more desirable than are beta compensations, and beta compensations are more desirable than are alpha compensations. That is, gamma compensations represent a more profound and deeper reorganization of cognitive structuring, which will therefore be more stable and resilient, than the other compensations. But this is exclusively an epistemological point. Experientially, we cannot say which compensation is preferable. Different compensations are appropriate in different contexts as we enact different political and social purposes. For instance, alpha compensation is the quintessential strategy of false-consciousness, but in denying a relation, it can also be a vehicle of fresh perception that discerns when the emperor has no clothes. Beta and gamma compensations are the strategies of advanced and integrated knowledge, but they also are utilized to sustain dogmatic thought.

Piaget also distinguished three cognitive loci at which equilibration may take place, in which the compensations are involved. These loci delimit a different range of phenomena than do the three compensations. In theory, the loci and compensations could generate a 3 x 3 matrix, though such a formulation would be oversimple. Compensations and loci are more fluid and multiply interconnected than a one-to-one representation would suggest.

First, there is equilibration between assimilation and accommodation as novel cognitive aliments are apperceived by the knower. The internalization and interiorization of Blake's poem, which jointly produce a reading of the poem, occurs in this locus. The degree of adequacy of the reading will depend to a large extent on the prior cognitive structures possessed by the knower into which the poem could be assimilated. Earlier I suggested that the internalization/interiorization process is a complex cognitive feat. Here I would briefly indicate some of the factors entailed in the single instance of internalizing and interiorizing Blake's short poem (or, to use Iser's terms, to construct its virtual dimension). Among factors influencing internalization, we might distinguish a knower's fluency as a reader (what are his basic skills of comprehension?), and his prior experience with literature in general (how closely does the knower read? are his habits of knowing tied more closely to oral patterns, emphasizing repetition, straightforward narrative, and illustrative digression, or has he internalized the skills of literacy, with heightened sensitivity to the structure and logic of narrative, to nuance <point of view, subtle shifts in meaning>, tone <irony, satire>, and figures <metaphor, metonomy, etc.> (Egan, 1987; Olson, 1980)? Factors which more properly impinge on interiorization would include what Jonathan Culler (1975), among others, have identified as the system of literary conventions which enable the attribution of meaning to texts. These would draw upon a knower's prior experience with poetry (does the reader understand the poem is to be read as a unity? that it should be read figuratively? that if read figuratively, should it be read allegorically or symbolically?). Interiorization also depends upon a reader's degree of cultural grounding (how familiar is the reader with Blake? with Blake's philosophical and religious mythology? how familiar with other Romantics? with the cultural context of early nineteenth century England?). Blake assumes an audience broadly educated in the liberal arts, that is, in the Biblical ("tents of Israel") and classical ("atoms of Democritus") traditions. Blake also assumes a general cultural familiarity with his own time. (Rousseau, Voltaire, and Newton are referred to in the poem. Will these associations resonate for the reader?) How at ease is the reader at constructing mental images (in the poem Blake relies strongly on an overall image of a desert encampment in a sandstorm)? And most importantly, what prior beliefs does he hold about, say, religion, or the value of reading a poem closely, or the status of science, that might color or prejudice his reading of Blake?

The varieties of ways in which the inadequate control of these factors may "deform" the text demonstrate as many varieties of alpha compensation. The protocols excerpted below offer a representative sampling. For example, Student 1, who responds,

> "The Scoffers" is about Israel and their long struggle to be an independent nation,

is reading the poem on a level of literality which expert readers understand to be inappropriate. But even a reading like that of Student 2,

> "The Scoffers" is about all those scientists and philosophers who mock God. No matter what they come up with, the fact that God exists and his chosen people are the Jews will never change. All their discoveries and explanations are just grains of sand compared with God,

is still too literal, assuming Blake's religious sense was strictly Judaic.

A particularly interesting reading is that of Student 3, who has reversed the sense of the poem in his internalization. He has framed Voltaire and Rousseau as heroes, rather than understanding that the poem is critical of them.

> "The Scoffers," written by William Blake, is an example of "Romantic poetry." The poem explains that the ideas of Voltaire, Rousseau, and other great men should be respected and not made fun of. Those who mock the ideas of these men are only hurting themselves because they do not benefit from the ideas these men have thought of and wrote about.

David Olson (1980) has argued that the written form of texts serves to preserve a kind of authorial authority within them; insofar as they are assumed to be archival, texts are not to be questioned or disagreed with. What I would suggest in this instance is that Student 3, a novice reader, misreads the poem in a way that preserves what might have been thought to be its authority. The implicit logic is something like: (1) all great men are to be respected; (2) the preservation of a person's words in print shows them to be great; (3)[by (1) and (2)], Blake is a great man; (4) therefore, those of whom Blake speaks, whose words have also been preserved, must be great as well. Misreadings of this type have little to do with a student's ability, but much to do with the progressive and time-consuming process through which the skills of literacy are internalized.

A reading which demonstrates beta compensation at the locus of assimilation and accommodation would be that of Student 4:

> Blake attacks Enlightenment leaders such as Voltaire and Rousseau for mocking society with progress. The sand they throw represents scientific advances. The sand shines upon Israel, suggesting its ultimate progress.

Student 4 has both incorporated the poem into his already integrated knowledge of the Enlightenment, reading it as a critique of Enlightenment rationalism, and demonstrated competence in using conventions of poetry, such as symbols and metaphors. I would suggest that Student 4's interiorization is something of a baseline paraphrase, the kind of reading which it is assumed a competent reader will formulate. As should be clear, this baseline level is already pitched at a degree of sophistication which implies mastery of a variety of literary conventions coupled with specific factual knowledge. The locus of assimilation and accommodation is the locus at which engagement with the text begins, and from which education may proceed.

At a second level, the equilibration process "has to do with interactions among the various subsystems of a total system" (Piaget, 1985, p. 7). Here our concern is with questions like, Do the same ideas show up in both the student's sense of "romantic" and in the poem? Can one or the other set of concepts be revised or modified to become compatible with the other? How conscious is the student of the implications of his ideas, that is, to what extent is there disequilibration (cf. Figure 2)? For instance, Student 1 offers an instance of alpha compensation at this locus. Having read the poem as Israel's struggle to be an independent nation (above), and describing "romantic" as

> to be full of thoughts of love ... caring ... . Loving and showing another person you love them plays a key role in romance,

he discounts the possibility of applying the description to the poem:

> "The Scoffers" does not fit my description of romance. This poem does not create a mood for romance. Although there are some bright lines which may lead to romance they do not. The poem does not create a feeling of love and openness which is needed in romance ... .

The opportunity to reread the poem, or to revise the sense of romance is simply denied or ignored, perhaps because both the reading and the sense of romance were too literal or naive to allow any cognitive interaction between them.

Conversely, in the protocol of Student 4, disequilibration is induced but reequilibration is problematic.

> *Description.* Many people use the term "romantic" to denote persons or things who are amorous. But the term has a more encompassing meaning. It describes anything concerning emotions, passions, or desires, relying not only on the concrete but on the imaginary as well. It denotes the full experience of an event and one's response to it.
>
> *Reading.* Blake attacks Enlightenment leaders such as Voltaire and Rousseau for mocking society with progress. The sand they throw represents scientific advances. The sand shines upon Israel, suggesting it's the ultimate progress.
>
> *Application.* My description of what it means to be romantic does not truly conform with the poem of Blake. It uses imagery and detailed description to create an optimal vision for the reader, but there is little in the way of emotions, or feelings that I described. This is not to say the poems are devoid of meaning or feeling, but that their meaning has a purpose other than creating an experience for the reader.

The initial description of romance is nicely differentiated, and the reading of the poem is adequate. Yet interestingly, the student resorts to an alpha compensation; he denies the relevance of the poem to his description when his rationale for the denial would seem to offer evidence for the opposite interpretation. The rationale itself consists of a series of backpedaling apologies. Student 4 argues that the poem creates an optimal vision, but without the right feelings; it induces the right feelings, but it doesn't create an experience for the reader. Yet it did create an optimal vision. The student seems to have argued himself in a circle, perhaps through having interpreted "romantic" too narrowly to mean only "romantic love." It is possible that Student 4 might remain sensitive to the tensions in his response, experiencing some discomfort or unease with his attempted resolution. This affective disequilibration could provoke cognitive reequilibration. For instance, one possible strategy would consist of incorporating a sense of awe shared by both romantic love and religious conviction into his sense of "romantic." A teacher might try to facilitate such a beta compensation, though whether or not cognitive reconstruction actually occurred would depend on the student.

Such a beta compensation is offered by Student 2.

> *Description*. "Romantic" is the catch all word for the expressions of one's feelings either toward another person, as in love, or about a particular event, such as a death of a close friend. In both cases, feelings are vividly expressed and easily seen.
>
> *Reading*. "The Scoffers" is about all those scientists and philosophers who mock God. No matter what they come up with, the fact that God exists and his chosen people are the Jews will never change. All their discoveries and explanations are just grains of sand compared with God.
>
> *Application*. My description of Romantic Poetry is too narrow. It can't be made to fit any of the poems because my description is restricting the word "Romantic" to love, feelings of the heart, relationships and so on.
>
> Romantic poetry is more than just the expression of love, but the expression of feelings of man, nature, and all other creatures of this world. Together with reason and emotions, Romantic Poetry shows how each creature is related and how one should care for each other and enjoy the world God has blessed us to live in.

Here an initial conception of "romantic" is expanded and differentiated through the disequilibration induced by the poem, to move from "expressive of feeling toward people" to "expressive of feelings toward people and non-people."

In Student 5, we see an example of gamma compensation at this locus. This student's initial description of romance is sufficiently differentiated for him to acknowledge disequilibration and then re-equilibrate by incorporating his reading of the poem within his description.

> *Description*. Romance is something that appeals to the emotion in a person, not the reason. A romantic uses their emotion to appeal to others' emotion. The sexual connotations of romantic also hold true to this definition. Candlelight dinners and moonlit walks on the beach are all appeals to emotion.

> *Reading.* He's saying that those who mock and scoff the Israelis will be punished eventually for their actions. The sand they throw will be blown back into their face.
>
> *Application.* "The Scoffers" is romantic because William Blake is expressing his emotion about the people who mock and scoff the Israelis. He tries to appeal to the emotion of the reader with his feelings. He certainly is not trying to reason with anyone.

Still it should be noted that while Student 5 employs gamma compensation, his ideas are neither sophisticated nor imaginative. I would reiterate that what is being described here is an epistemic activity; it bears no necessary relation to a given level of interpretive quality.

A third locus of equilibration is entailed in the hierarchal integration of "lower level" cognitive structures into a "higher level" component of the total cognitive system of the knower, in which the lower level structures serve as differentiating factors. In the example we are working with, questions such as the following would be implicated: How do poems (or literature, or art) exemplify ideas? Mimetically? Allegorically? Symbolically?; or What is the grounding for how I think about ideas, e.g., how is a 19th century poem relevant to my ideas about being romantic or to Romanticism as an intellectual movement?; or Do I agree with Blake? Am I religious? Do I condemn science? Am I romantic? In what ways? To what extent? As a romantic, do I hold the same beliefs and attitudes, and in the same way, that the Romantics did? Questions of how new ways of thinking about romantics stand with respect to the pre-existing and potentially deeply held beliefs of the knower are raised. What cognitive behaviors can we expect when belief systems are engaged that are constitutive for the knower's identity, that is, when personal schemas come into play? The short answer is that these schemas may either enhance or retard integration of novel aliments in a variety of ways by modifying their construal. Let me illustrate this through three more protocols.

In this protocol of Student 6, we see behavior much like that of Student 4 in that application is resisted. On the other hand, we are made aware of the personal concern motivating this resistance.

> *Description.* I define romantic as it applies to situations rather than specifically in the sense of romance in poetry. Something is romantic if it has a very special significance to Amy and I, something that we had done before together, so something that we had been hoping to do. When we are alone together, sharing, loving, enjoying each other's company. Something becomes romantic

> because of the mood, or the time, or just the fact that we are together.
>
> *Reading.* It doesn't matter what Voltaire and Rousseau say, it is worthless because their accusations turn around on themselves. Every accusation they make is all part of the divine, hurting the doubter and glorifying Israel. All of the science of Democritus and Newton is nothing compared to the truth of God.
>
> *Application.* My description does not really fit. Romance in poetry attempts to take an event and glorify it and cast a particular beauty on it. The poem does share an appeal to the emotions, but my definition is more specifically concerned with appealing to a special love that already exists between two people. Romance in poetry does not necessarily appeal to love so much as thought and appreciation and beauty.

Disequilibration is induced but reequilibration entails introducing an epistemologically dubious differentiation. The student recognizes the poem and description have relevance for one another, but seems to not want to coalesce the two. He chooses to differentiate meanings of "romantic," and thereby preserve what is apparently the special importance which romance has in connection with his girlfriend. Given, however, that the series of "thought and appreciation and beauty" used to describe romance-in-poetry arguably informs his conception of romance-as-love as well, it is quite possible that but for the personal factor of his relation with Amy, the two conceptions could have been brought into closer relation.

Student 7 offers a description of romance and of readings of "Mock on, Mock on, Voltaire, Rousseau," and also of a second poem by Blake, "The New Jersualem" (see Appendix III), strongly influenced by his orthodox Christianity. It offers an example of different compensations at two different loci. It also raises issues of the interaction between epistemology, pedagogy, and personal ideals.

> *Description.* To be romantic is to be in love, caring, thoughtful, sensitive, warm, affectionate, open-minded and having general overall good feelings about all that is around you and the person you are in love with .... People who are romantic should also conduct their actions in just ways as the Lord has established, not in ways of sin. I believe to experience the essence of being romantic, the people should do all that they enjoy the most, experience those feelings

stated above, and do all that they do in the glory of God's works.

Reading of "Mock on, Mock on, Voltaire Rousseau." Blake is saying that people who question, criticize, satire (sic) or try to discover hidden meanings behind all on the earth are making no progress towards eternal life. Their ideas are being blown up in their faces and nothing is accomplished really or any truth found from their work in God's sight. Blake believes that the Lord has made everything his own way, and it is not right for humans to question his works or define his creations because no one knows specifically what the Lord means all throughout his earth.

Reading of "The New Jerusalem." Blake is questioning (arbitrarily) whether Jerusalem was destined to be built in England and not in Israel, and whether Jesus was destined to walk England's lands and receive God's shining glories there. If this is what was truly destined to happen, then Blake would use all that he owns and all his power without sleep in order to build and protect Jerusalem at all costs. I think he believes that if he did this, then the people of England would have been the chosen race, Jesus would have been saved and not crucified on the cross, and that all the glories of England's countryside would be proclaimed as the Promised Land.

Application. The formal definition of romance is a story that tells of heroic deed, adventure, and love. My definition of romance deals more with the affection side of love, rather than adventure and being a hero. A romantic poem includes both the formal definition and my interpretation too. William Blake's poem on Jerusalem applies to the formal definition very well. Blake is showing the heroism aspect of romance by being willing to fight for this country and Jerusalem if it was built there. He also shows his love for God, Jesus, and his country through the way he describes the beauty of England and through being willing to do anything for his country to protect it ... . My definition could be made to fit the poem if I concentrated more on the adventure and heroic deeds aspect of romance, rather than all love and affection.

The personal schema here is evident, as is its influence on the reading of the poems. Student 7 has assimilated these poems deeply, seeing them as affirmations of what he already believes. This exemplifies gamma compensation at the locus of integration, but at the locus of assimilation/accommodation, Student 7 seems to engage in alpha compensation. In terms of what we know of Blake, Student 7's readings are superficially misreadings, but on a deeper level perhaps they are creative misprisions. "Mock on, Mock on, Voltaire, Rousseau" does not at first seem to be about "progress towards eternal life," or the impropriety of investigating the creation, yet in terms of Blake's larger mythology, maybe it is. Blake, after all, used the myth of the fall (which, among other things, seems to be about the price of moral knowledge) to symbolize his own sense that man had fallen due to a fragmentation of psyche. Is not "progress toward eternal life" a possible trope for reintegration?

It is true that the reading of "The New Jerusalem" in terms of the Old Testament sacred-history would seem to be more contrived, that Student 7's reading of Blake's clarion call as a literal call to arms is implausible. And perhaps his misprision of "Mock on, Mock on Voltaire, Rousseau" is not creative after all, but just a misreading. I, as the teacher, may want to help this student enrich his understanding of Blake and his reading of poetry, that is, to better master the interpretive conventions that further internalization. But even then his unbridled religiosity is likely to color his readings. The fact that he is fully aware of the passionate spirituality of Blake, and that these poems live for him, are already among the most desirable of educational outcomes, outcomes that could be jeopardized by a heavy handed assertion of pedagogic authority. In this particular instance, there would need to be a full social negotiation that well might include others (other students, other faculty, Blake scholars, other devoutly Christian readers, etc.) before his reading was judged appropriate or not. In such a way arises the authority of interpretive communities (Fish, 1980). Student 7 instantiates both the epistemic complexity of cognitive activity occurring on several loci, and the necessity of the social negotiation of interpretive meaning.

In Student 8, we see an example of beta compensation at this locus, where personal schemas come into play even at the level of description. What is noteworthy in this protocol is the degree of specificity evidenced from the start. Explicit personal recollections are tied to more general discussions, and in fact dominate them.

> *Description*. If someone is to be considered romantic they must be able to stir some sort of emotional response.... . The response may take on many forms. A sense of great joy, intriguing adventure and/or extreme duress can all result from a romantic endeavor. These emotional responses are brought on through the whims of the person

and can not always be controlled through conscious thought ....

One day I saw a girl who was obviously disturbed by something. She looked as if she was going to cry. No one but me seemed to notice or care. I approached the stranger and she told me her horse had a miscarriage. It was very ill and might die. With a little effort I managed to raise her spirits. She produced a teary smile. I will never completely understand why, but through the years I went to great pains to make her happy. One summer I "hitch-hiked" from Watertown to Morrisville to see her at her ranch where she trained horses. It took me all day and night to get there. A good majority of my time was spent walking in the cold clear night, which I was unprepared for ....

Reading. Your mocking of Israel is further blinding yourself. Israel will remain powerful because nature and all the elements are on its side.

Application. My previous description of what it means to be romantic ... left out the idea that romantic feelings can appear in emotional responses to an event in one's life, even if it did not wholly relate to another person ... . The following is a romantic encounter during solitude.

It was early May when I had arrived in Alaska, laden with an overstuffed backpack and a heavy box of dried food. It was cold on the beach, but there I stayed because the opportunity for employment was greatest there. It was not long before I was completely out of food and went day to day bundled in my clothing, continually feeling the bite of hunger. I eventually found a respectable ridge of sand and stone where I could pitch my tent and be relatively free of the cold, direct wind off the ocean. One morning I woke and peered out of my tent. The wind was still and a bright sun warmed my tent to a comfortable temperature. That morning I became completely absorbed in the supreme serenity and beauty of the world around me. The rugged snow capped peaks all around, the shimmering water and everything else around me took on one shape. For those few hours I was gifted with complete inner peace. No longer did I have regrets for bringing myself to this infinite and unforgiving place.

This experience is consistent with the idea that a romantic moment is an emotional one. I usually think of the romantic moment as being favorable or desirable. The

> moment becomes more intense when surrounded by difficulties or hardships.
>
> My description of what it means to be romantic has some bearing on the poem. The poem is an expression of an extreme emotional response to some event in life. In this sense the definition can apply. The poem appears to be making a statement about unhappy things in life or dilemmas not resolved. The idea that romantic writing contains unpleasant concepts is not at odds with my description. But these poems contain very little pleasant verses. This puts my definition at odds with the poetry since some sort of pleasure is required in a romantic situation.
>
> The description can be altered to better fit the poem. We must leave out the idea that being romantic is central of the feeling of love, desire or other pleasant thoughts. We can be settled with the idea that being romantic is centered around all emotional responses. They can be pleasant or unpleasant. Poetry is romantic because it is descriptive of emotional response.

Aside from the clear personal memories informing his account, this student is also noteworthy for his cognitive flexibility. As his convictions are clearly grounded in explicit, detailed recollection of experience, so it would seem he can demonstrate the resilience of entertaining new perspectives, changing his ideas, of engaging in beta compensations. In this case, the recollections are not jeopardized; they do not change. Because they are so well differentiated, they maintain their integrity even as the convictions they inform are altered. In other cases at which we have looked, for instance that of Student 4 and Student 6, we have seen a curious resistance to cognitive revision (to beta compensation) even when the inertia of their logic pulled them in that direction. I would argue that their resistance follows from the relatively undifferentiated form in which personal schemas grounded their belief, just as in the opposite case of Student 8, his resilience goes hand in hand with his high differentiation.

My theoretical intent in presenting these protocols has been to demonstrate how personal schemas held by students - in this case, what they think is meant by "romantic," and the varieties of different experience that informs their meaning - become operative in their appropriation of subject matter - in this case, a Romantic poem as a propaedeutic to the larger issue of Romanticism. The pedagogical intent is to identify a handful of typical responses suggesting different levels of engagement, which themselves point toward the kind of instructor interventions that might be facilitative in the classroom. I hope I have been able to show that apart from the problems that make classroom learning difficult - which I have here collectively referred to as the problems of literacy - an entirely separate set of concerns, having to do with personal schemas, must also be taken into account. A fundamental task we have as teachers is eliciting students' personal schemas, and then, as they are elicited, helping students to fully examine them, to see what dimensions of personal experience legitimize them. Education as *paideia* in this sense is also emancipatory.

As we have seen, personal schemas tend to be intuitive and holistic. They are neither as conceptually precise as ideas, nor as systematic as metaphors, but incorporate both ideational and metaphoric components. They lack the specific operational reversibility of cognitive structures. In function, they resemble the cognitive schemas which inform much current work in script theory and story grammars. Yet since they include a dimension of personal relevance, their closer affiliation is with the affective rather than with the purely cognitive. It is this personal dimension that requires that they be understood hermeneutically, not deductively. As general organizing devices, their educational function is central; they, and not anything exclusively resident in either a text or a teacher, provide the means of access through which texts and classroom instruction may resonate for the student. In turn, they only come into play in the social nexus where a knower is asked to engage an artifact.

An Integrated Guiding Image

The examples looked at above concern student responses to teacher-generated questions. Students were constrained to think about topics and produce responses in connection with an introductory assignment for which they may have had no initial interest. While I affirm the value of such work, and of my pedagogy in approaching it through students' naive conceptions, I hasten to add that this imposed task performance is only a first step toward what is meant by education as *paideia*. Clearly, a number of moments can be identified in the ongoing process of constructing meaning, and appropriating that meaning in counterpoint to the concerns of the self. As should also be clear, there are a number of different strategies which students adopt as they appropriate texts, and not all of these will be fruitful.

Let me move on to the perspective introduced at the beginning of this essay, to that of insights offered by students on their own initiative. These latter insights are not introductory; rather, as final summations for the semester, they represent some degree of integration of subject matter. It is here that personal schemas will have interacted with the text over time in such a way that certain images from the text will have become centrally involved in the students' appropriation. This is part of my claim concerning personal schemas, but my claim goes further.

Earlier I said that such a schema is evident only some of the time. What happens when it is not? When no personal schema is evident in interviews, discussion of the text is invariably *pro forma*; the student is likely to remember, accurately or not, some number of details from the book or from class discussion of it, but recollection will not show any coherence or be organized in a way that demonstrates the text will have a lasting significance for the student. Rather, what emerges is a sort of random recall, done primarily to please the instructor. In a short time it can be expected that the recollection of the text (and therefore its potential value for *paideia*) will have diminished all the more. My further claim, then, is that unless personal schemas are engaged by the text, offering an objective correlative for aspects of the knower's sense of personal identity, there will not be lasting appropriation of the text. Rather, there will be at best only the rote behaviors - memorization and regurgitation on demand - which connote internalization but not interiorization.

Let me offer a single example in which personal schemas were engaged. The text in question is the *Philoctetes* of Sophocles.[1] Student 9's transcript is exceptional in the succinctness with which it allows us to see a personal schema come to clarity in interaction with a text, but it is typical in its exemplification of the construction of meaning. (The full transcript of this student, Student 9, is reproduced in Appendix IV.)

Student 9 is first asked to state what he sees as the central theme of the play (part of internalization/interiorization).

> S: The theme was that, in my opinion, that this was the point where the Greeks started to wonder about their own *logos*,[2] because characters changed their, uh, but not the main characters, but the characters around the main characters changed their friends to enemies and enemies to friends, the change which is the ultimate bad thing in Greek literature, but they did it and it seemed like the right thing to do, which makes one wonder what was really right.

The student has interiorized both the play and a set of concepts (*logos*, Homeric morality, *arete*, honor, etc.) used to develop the Greek world view. Thus, he is able to identify a clear tension in the play in terms of these concepts.

> P: OK. Do you remember any specific things, any specific scenes or events or whatever?
>
> S: Well, one was when Neoptolemous finally decides that he's not going to listen to Odysseus any more, and that he's going back to Philoctetes and giving him back his bow. Right there was the turning point. That's the one thing you wouldn't expect to happen.

P: Why not?
S: Because from most of the Greek that we read, they absolutely could not change if they wanted to be honorable, and he did change and he became honorable, which is interesting.

He understands the importance of honor, but sees that, ironically, in the Philoctetes it is Neoptolemus's act of generosity toward Philoctetes, which is simultaneously a betrayal and disgracing of Odysseus and the Greeks, which will win him honor. In response to a probe, the student then relates the problem of honor to our own time.

P: That's good, that's good. So, um, ... is honor important?
S: It was to the Greeks. That's what they had to shoot for. Success was honor. That's the only thing. I mean being wealthy was one thing, but if you weren't honorable, it wasn't worth it, and besides that, everything surrounded honor because the gods looked for the most honorable, and so on.
P: OK. Honor doesn't play that role for us, does it?
S: No, it doesn't, because society today is somewhat warped from what any society should be.

In response to a probe ("So, is our society more corrupt, do you think, than the Greek culture?"), Student 9's response is initially an alpha compensation. He wishes to avoid being forced to judge and condemn his own time period. But his rhetoric betrays him: we are "a little more corrupt."

S: It's not more so. It's just that it's not regarded as bad now, it seems to be, it has a tendency to be a little more corrupt cause more people don't care if they are honorable or not. Society's not shooting to be honorable. Instead, they, you only live once, and so on.

Yet this is a fortunate admission, for it allows Student 9 the opportunity to clarify his difference from his own time -

P: Do you think it's important to be honorable, you personally?
S: I do, yeah ... . I feel if I'm not doing the right thing, I won't do the personal struggle for honor:

- and to ground this difference in the play by identifying with the one character who best exemplifies the personal struggle for honor:

P: Of the characters in both plays - Ajax, Menelaus, Agamemnon, Telamon, Neoptolemus, Philoctetes, Odysseus - you know, who do you kind of like the most, who do you admire the most?

S: The most. Whew. I would have to say Neoptolemus, because he did change, because he was expected to listen to Odysseus because that's their heritage, they listened to their elders, and if their elders told them that something was right, then they did it. But he decided for himself that something else was right. He went and did his own feelings over tradition.

Note that if honor alone, as a global category, were his only concern, a number of characters appropriately could have been selected - Philoctetes, Ajax, Odysseus as he behaves in *Ajax*. But his identification is precise. Neoptolemus is his objective correlative. This becomes clear as I try to disequilibrate him by suggesting a possible inconsistency in his remarks.

P: How would you respond to the following kind of argument that said, well here, you're telling me that you admire Neoptolemus, but he changed his mind and went with his feelings, but a few moments ago you said that that was what was wrong with our culture, that people just did what they thought was right.

S: Well, he went with his own feelings to do what he thought was honorable. He was shooting for honor, he wasn't shooting to improve himself, I mean if he was shooting for his own improvement, then he would have helped Odysseus fool Philoctetes, but instead he decided on his own that the right thing to do, that he decided for the right thing instead of for his own thing. Basically.

The student's response preserves and clarifies the differentiation, separating Neoptolemus from honorable elders ("deciding on his own"), and from what the student sees as the self-interest of today ("he decided for the right thing instead of his own thing").

My argument has been that cognitive construction in the humanities does not overcome or sediment the figurative, but functions in counterpoint to it. This epistemic activity is like Winnicott's transitional object in that it occurs on the interface where self meets world, where the boundaries between external and internal are blurred. The text provides a variety of thematic images, some of which will be utilized by the knower as he or she constructs meaning. The figuration provided through the text is negotiated in the interiorizations undertaken by the epistemic subject as it equilibrates with respect to the text. Usually these epistemic negotiations are socially mediated, either in the classroom or in informal discussion. The knowing that results is not necessarily decentered; it does not always end with an objective coordination of perspectives. And while it typically does entail cognitive reequilibration, such equilibration may be minimal and superficial. Still, such knowing can be richly contextualized and empowering, education as *paideia*, a knowing in which the domain of self-understanding is itself differentiated through the act of appropriating the text.

Let me end by situating our role as teachers with respect to the preceding discussion. We are neither ideologues, pushing a party line or single interpretation, nor are we divine knowers to whom a singular dispensation has been granted. We are fallible learners, just like our students, bringing our own personal schemas to bear as we too seek to appropriate what artifacts afford. (My own limitations are brought home to me when I encounter students like Student 7, whose religious perspective I do not share.) And just as I find the reasons students resonate to particular texts both surprising and legitimate, so in education generally we mutually learn from each other, negotiating the construction of meaning in what has been variously termed "paradigms of inquiry," "the conversation of mankind," "interpretive communities," etc. It seems to me that among our chief tasks as teachers is to facilitate for our students the process through which they and we jointly seek clarity. This done, content knowledge of the kind Hirsch (1987) has associated with cultural literacy will follow of its own accord. This not done, the knower will have no reason to bother looking for it. In the humanities, education has less to do with learning technologies and the disciplinary distribution of information than with personal relations, with what both teachers and students bring to the classroom, and are willing to share there. Is education in mathematics so different?

Notes

1. Sophocles' *Philoctetes* is set in the last years of the Trojan War. Following Ajax's suicide, an oracle reveals that the Greeks will be unable to capture Troy unless they are aided by the bow of Heracles, which is in the possession of Heracles' former hunting companion, Philoctetes.

   On the way to Troy, Philoctetes had been bitten by a serpent (some say as punishment for sacrilege against a god). The wound festered and gave off such obnoxious odors that Philoctetes was set ashore on the island of Lemnos and abandoned there.

   Following the oracle, the Greeks realize they must persuade Philoctetes to relinquish Heracles' bow. They send an embassy headed by Odysseus and, in the version of the myth used by

Sophocles, the son of Achilleus, Neoptolemus. The main action of the play concerns the moral struggle of Neoptolemus over his conflicting loyalties: between loyalty to the Greeks and Odysseus, who would have Neoptolemus use treachery to recover the bow, and loyalty to Philoctetes, with whom Neoptolemus forms a friendship but who refuses to aid the Greeks.

2. The *logos* is, roughly, the deep overall "logic" of the universe, what makes it ordered, a "cosmos." The *logos* orders the affairs of both the physical universe (nature) and the affairs of human beings (human nature). Thus the *logos* connected humans to the larger "logic" of the universe.

Appendix I

LS 196/Lewin
For Monday, February 8, please do the following written assignment:

1. What does it mean to be "romantic?" Write a statement (1-2 paragraphs) on what it would mean to describe someone or some event as "romantic." Give at least two concrete examples to illustrate your description.
2. Read the poems from Williams, ed., *Immortal Poems* listed below, all of which are conventionally understood to be examples of "Romantic poetry." Write a 2-3 sentence paraphrase of "The Scoffers." In addition, pick a second poem and write a brief paraphrase of it also (2-3 sentences).
3. Try to apply your description of what it means to be romantic to the poems. Can they be made to fit the poems? If so, how? If not, why not? (1-2 paragraphs.)

Please read:

William Blake,	"Auguries of Innocence," p. 227 (only)
	"The Scoffers," p. 232
	"The New Jerusalem," p. 233
William Wordsworth,	"My Heart Leaps Up," p. 251
	"Composed Upon Westminster Bridge," p. 252
	"The World is Too Much With Us," p. 260
	"It is a Beauteous Evening," p. 252
John Keats,	"Ode on a Grecian Urn," p. 321

Appendix II

Mock on, mock on, Voltaire, Rousseau,
    Mock on, mock on; 'tis all in vain;
You throw the sand against the wind
    And the wind blows it back again.

And every sand becomes a gem
    Reflected in the beams divine;
Blown back, they blind the mocking eye,
    But still in Israel's paths they shine.

The atoms of Democritus
    And Newton's particles of light
Are sands upon the Red Sea shore,
    Where Israel's tents do shine so bright.

-William Blake

Appendix III

The New Jerusalem

And did those feet in ancient time
    Walk upon England's mountains green?
And was the holy Lamb of God
    On England's pleasant pastures seen?

And did the Countenance Divine
    Shine forth upon our clouded hills?
And was Jerusalem builded here
    Among these dark Satanic Mills?

Bring me my bow of burning gold!
    Bring me my arrows of desire!
Bring me my spear! O clouds, unfold!
    Bring me my chariot of fire!

I will not cease from mental fight,
    Nor shall my sword sleep in my hand
Till we have built Jerusalem
    In England's green and pleasant land.

-William Blake

Appendix IV

Protocol of Student 9

P: We're going to talk about Sophocles' plays, the *Philoctetes* and the *Ajax*. So, how come you want to talk about them?

S: Well, because I worked a lot, looked into them twice now, right down to the fine points, that's probably what I know the best.

P: Did you like them?

S: Yeah, I did. I liked both of them, cause...they both had a good theme.

P: What was the theme?

S: The theme was that, in my opinion, that this was the point where the Greeks started to wonder about their own *logos*, because characters changed their, uh, but not the main characters, but the characters around the main characters changed their friends to enemies and enemies to friends, the change which is the ultimate bad thing in Greek literature, but they did it and it seemed like the right thing to do, which makes one wonder what was really right.

P: OK. Do you remember any specific things, any specific scenes or events or whatever?

S: Well, one was when Neoptolemus finally decides that he's not going to listen to Odysseus any more, and that he's going back to Philoctetes and giving him back his bow. Right there was the turning point. That's the one thing you wouldn't expect to happen.

P: Why not?

S: Because from most of the Greek that we read, they absolutely could not change if they wanted to be honorable, and he did change and he became honorable, which is interesting.

P: That's good, that's good. So, um,...is honor important?

S: It was to the Greeks. That's what they had to shoot for. Success was honor. That's the only thing. I mean being wealthy was one thing, but if you weren't honorable, it wasn't worth it, and besides that, everything surrounded honor because the gods looked for the most honorable, and so on.

P: OK. Honor doesn't play that role for us, does it?

S: No, it doesn't, because society today is somewhat warped from what any society should be.

P: Oh, how come?

S: Cause of all the crazy people in the world, because of the lack of importance of religion....and people now are just out for themselves mostly....My favorite thing to talk about is the religious scandals, the television evangelists....

P: So, is our society more corrupt, do you think, than the Greek culture?

S: It's not more so. It's just that it's not regarded as bad now, it seems to be, it has a tendency to be a little more corrupt cause more people don't care if they are honorable or not. Society's not shooting to be honorable. Instead, they, you only live once, and so on.

P: Do you think it's important to be honorable, you personally?

S: I do, yeah....I feel if I'm not doing the right thing, I won't do it.

P: Of the characters in both plays - Ajax, Menelaus, Agamemnon, Telamon, Neoptolemus, Philoctetes, Odysseus - you know, who do you kind of like the most, who do you admire the most?

S: The most. Whew. I would have to say Neoptolemus, because he did change, because he was expected to listen to Odysseus because that's their heritage, they listened to their elders, and if their elders told them that something was right, then they did it. But he decided for himself that something else was right. He went and did his own feelings over tradition.

P: How would you respond to the following kind of argument that said, well here, you're telling me that you admire Neoptolemus, but he changed his mind and went with his feelings, but a few moments ago you said that that was what was wrong with our culture, that people just did what they thought was right.

S: Well, he went with his own feelings to do what he thought was honorable. He was shooting for honor, he wasn't shooting to improve himself, I mean if he was shooting for his own improvement, then he would have helped Odysseus fool Philoctetes, but instead he decided on his own that the right thing to do, that he decided for the right thing instead of for his own thing. Basically.

P: OK. Let me ask you to think about a middle ground....

11 Enhancing School Mathematical Experience Through Constructive Computing Activity

Larry L. Hatfield

Within the literature of mathematical education, one can find little explicit attention to students' mathematical experiences. Perhaps this is due to the complexity, indeed impossibility, of adequately characterizing what constitutes these experiences.

Our literature and professional meetings are replete with personalized advocacies and admonitions about how one should teach mathematics, what ideas from mathematics should be included in school mathematics, or what classroom activities should be provided to our mathematics students. From these expressions, one can often infer important qualities of what students *might* experience in the mathematics classroom. These expressed viewpoints deal with possible or potential mathematical experiences or with idealized notions of expected or recalled "realities." But, we educators must realize and accept the ultimate dilemma of our role: experience, the essence of teaching and learning, cannot be directly or explicitly known or shared.

My interpretation of the goal of this conference is that we are seeking to advance our thinking about the nature of human experience in that domain we call mathematics and to focus especially upon the nature of students' experiences in learning mathematics. My purpose in this paper is to attempt to share some aspects of my own views of higher quality mathematical experiences to be stimulated and supported through effective computer usages by students and teachers in the classroom.

I feel the need to point out several precautions, disclaimers, and limitations in my own attempt to contribute to the conference goal:

1. Surely, we must be conscious of the variety and complexity of hidden assumptions we each may be making about the nature of *mathematics as a human activity*. As we seek to analyze and discuss the construct of mathematical experience, the meanings, emotions, and beliefs we each have about "mathematics" will surely differ. These differences may be at the root of disagreements and confusions we may have with the views apparently expressed by others. But the processes of pursuing our own realized

understanding of "mathematical experience" will probably force us to sharpen our thinking about the nature of mathematics as we know it.

2. We also will have differing assumptions about the purposes, contents, and consequences of *school mathematics*. What should determine the nature of school mathematics, as intended by a planned curriculum? Here, we each have probably constructed some more or less specific interpretation of the direction and substance of some idealized experiences students might have.
3. These personal interpretations of "mathematics" and "school mathematics" reflect the specific experiences we have had in learning (and for some of us, in teaching) mathematics. None of us would claim to know or even possess vague conceptions of the totality of recorded information that is mathematical! The vastness of what humans know or knew that is mathematical is beyond knowing. Yet, we each claim to know enough to be able to address the complexities of the topic of focus here. But what is the basis of our claim? Ultimately, what I think about the human mathematical experience is based upon only one human's experience---*my own*.
4. I have been a student of mathematics for a long time. Indeed, I might assert that I have experienced mathematics throughout my life, but to do so would require that I make much clearer what constitutes "a mathematical experience." Even when I reflect upon what I can remember of my formal schooling period, today I prefer not to accept as "mathematical experience" most of the daily classroom situations I experienced. I level a similar implied criticism about much of the experience I perceive today's students are having in mathematics classrooms. This criticism is surely the result of my current biases, expectations, and beliefs.
5. From my experiences of teaching mathematics and using computers in my teaching of mathematics to others, I have another vast collection of remembered experiences. These memory fragments of the occasions and events which I shared with students serve as a primary source of my current ideas about the topic of this paper. Central to my concern here is what I remember now about my perceptions and inferences of what my students appeared to be experiencing during those past occasions. These memories will be a primary source of my ideas for enhancing mathematical experience through constructive computing activity.

I hope this "pre-ramble" has not totally undermined the reader's acceptance of my intention to deal with some of my viewpoints on the nature of school mathematical experience. But I wish the reader to realize that, from my constructivist perspective, I acknowledge the impossibility of knowing with certainty anything at all about the mathematical experiences of others and the tremendous difficulty of even being clear or unbiased about my own past mathematical experiences.

Aspects of Significant School Mathematics Experiences

Plunkett (1981) has presented a lucid argument for abandoning the question "What is mathematics?" since it presumes a realist's objective existence of a thing called mathematics. He poses the question "What (mathematical) is going on in a mathematics classroom?" as one which may be possible to answer. Again we are stuck with the inevitable inferences to be made when we attempt to conclude something about a person's experience from the individual or group activity we perceive and interpret. However, one possible outcome of such analyses of what we infer may be "going on" in a classroom may be the articulation of differing notions of what "doing mathematics" means. This can lead to a search for the underlying reasons for the divergent views we might hold, and this might reveal important qualities of mathematical experience that we understand and value.

To me, mathematics students *do* many different things which appear to have elements of important mathematical experience. Beyond the observable actions of listening, looking, drawing, writing, asking, or stating, they seem to engage in a complex array of important mental activities as they attempt to cope and function within the situations posed by the mathematics teacher or textbook. These mental activities may include some aspects of such hypothesized intellective and emotive processes as:

> abstracting, particularizing, simplifying, generalizing, assuming, defining, conjecturing, guessing, checking, denying, confirming, concluding, questioning, doubting, justifying, proving, remembering, applying, classifying, connecting, matching, visualizing, computing, estimating, differentiating, patterning, valuing, fearing, hating, risking, and complying.

Such a list of categories could be extended even further and might be made more useful through an attempt to characterize what might be involved in each. But let me contend that rich mathematical experience must necessarily involve complex assemblies of such processes in ways which not only engage, but draw attention to the very nature of such processes. In this sense "mathematical knowledge is a structure of thinking---the structure is a structure of processes" (Thompson, 1985, p. 192).

A metaphor which I like to use might be helpful. I sometimes think of such intellective processes as the threads in the tapestry of a person's "mathematics." Of course, each person is ultimately the weaver of his or her own tapestry. Each tapestry is unique and its complexities and subtleties prohibit even the weaver from ever seeing it all clearly. The tapestry is never finished. For most humans the tapestry begins very early in life and is woven most intensively and purposefully during the years of school mathematics. But how this weaving occurs and the beauty, strength, and utility of the tapestry depends upon the nature and circumstances of the mathematical experiences. These threads are both produced and subsequently used by the weaver. As educators we cannot provide the "threads" or do the "weaving" for the student (though many teachers seem to proceed as though they want to or believe they can), but I assume that we can greatly affect the nature and quality of the threads which comprise the student's tapestry and of the weaving that produces it.

In most school mathematics classrooms that I visit and observe, I am discouraged by much of what I think is occurring in relation to the threads and the tapestry as I believe they might exist in students' mathematical knowledge. It is not that these threads are necessarily totally missing; indeed, given the goals and concepts which seem to be present in school mathematics, it would be impossible to prevent the participants from experiencing some aspects of these processes, these "threads," in their daily activity. Since these intellective processes appear to be integral to human thought, and mathematics is pre-eminently an activity of the human mind, it may be impossible under any form of mathematics instruction to eliminate such cognitive processing. What appears to be generally lacking is a *quality* encounter with the situations and ideas that stimulate student activity in which an individual's process "threads" develop and are used[1].

What constitutes a high quality mathematical experience? What are some of the qualities I would seek in the mathematical encounters of students? Following the suggestion made by Mason (1982), I would organize these as two distinguishable but inter-related states: *emotional states* and *enquiry states*. By focusing upon the states of feeling and thinking of the student as the primary determinant of experiential quality, I want to emphasize a shift of focus from environmental or situational factors to the student's status and viewpoint. Mathematics teachers who become much more concerned with trying to understand these internal conditions may be able to enhance the mathematical experiences of their students.

Fortunately, these conditions internal to our students appear to be influenced to various extents by factors directly accessible to the mathematics teacher. However, I suspect that these conditions within the student are predominately influenced by the nature of the person's mathematical "tapestry" presently constructed. The effect of past experience is such a powerful precondition, that it becomes exceedingly important to nurture quality mathematical experiences early and at every stage of the child's schooling. Though it may be possible to overcome some consequences of an experiential history of poor quality encounters in school mathematics (in much the same ways that the cumulative detrimental effects of prolonged malnutrition might be somewhat reversed), it is likely that the irreversible damage and limitation would continue to be revealed in the student's mathematical activities. The mathematically "starved" may be doomed to severe handicaps which permanently limit their potential for significant intellectual functioning and growth.

Emotional States

The first domain of conditions for quality mathematical experience involves the *emotional states* of the student. What can we, for whom mathematics has been, and continues to be, an aspect of our existence, recall of our own emotions which relate to remembered mathematical experiences? As we consider such a question, we might identify such descriptors as challenged, enthusiastic, proud, interested, curious, playful, and inspired, and such emotive terms about "mathematics" as beautiful, powerful, intuitive, massive, useful, ideal, intriguing, rigorous, structured, consistent, and open. We might also admit to having experienced emotions and feelings which could be described as frustrated, overwhelmed, confused, fearful, and bored, or with adjectives such as risky, abstract, uncertain, mundane, demanding, irrelevant, inflexible, formal, arbitrary, technical, picayune, and vague for "mathematics."

In other words, we probably all can make connections to both negative and positive emotions in our recalled mathematical experiences, and these would probably not be unlike those which would be expressed by today's mathematics students. Overall, we would probably differ from school mathematics students in the degree or extent of significance we attach to our positive versus negative emotions. To promote higher quality mathematical experiences, we would want the emotional balance of experience to tip toward the side which reflects the type of positive descriptors suggested above. That is, the emotional elements in the student's mathematical experience are significant, albeit subtle, aspects and must be considered as such.

One of the most important emotional elements involves the *sense of purpose* in approaching and pursuing any mathematical situation. Surely, this awareness of purpose may often be rather vague and unclear in the early stages of encountering a particular situation, but it should be an element of concern throughout, with the realized purposes becoming progressively clearer as the nature of the situation and its possible outcomes become clearer. Without actualizing adequate purposes which relate to important "meaning-givers" in the student's life, the consequences of experience are likely to be lessened. Productive mathematical activity must be purposeful, and the key factor must reside in the intention of the student.

It appears to me that in the culture of the typical school mathematics classroom both teachers and students seem to proceed without sufficient concern for developing, clarifying or reinforcing purposes for much of the activity experienced. To me, effective mathematics teaching stimulates and guides students to construct not only understandings of mathematical ideas but also recognition and appreciation of the importance of mathematical thinking upon one's intellectual growth and development (for this is perhaps the most significant purpose for being a mathematics student). Guidance can be given to mathematics students aimed at promoting the constellation of positive intentions that may be advantageously associated with constructive approaches to learning.

Students should be encouraged to ask (at least of themselves) such questions as: What am I going to be doing? Why am I embarking on the task before me? What might I find out from my activity? Where does this situation appear to fit-in to what I have experienced before? Why am I being asked to do these things? How might my efforts and results in this activity be of value to me now and in the future? Instead, it is only rarely (and perhaps for negative reasons) that an occasional student will ask: Why do I have to do this? And most school mathematics teachers seem to struggle in their attempts to provide an adequate answer. On the other hand, it may be that analyzing one's purposes too extensively may lead to paralysis. The pleasure of constructing mathematical concepts or solving mathematical problems might best be left as the experience, per se. However, I would suggest that most of the time school mathematics students are constructing their mathematics with personal goals and purposes which do not connect at all with higher aims of intellectual growth and development, being focused instead upon such matters as getting a score or grade; avoiding rebuke of teacher, parent, or peer; or complying with adult-imposed (both present and future) schooling requirements. "All is lost upon him who has no felt purpose for being."

Another conditional element in the domain of emotional states relates to the student's self-perception of *potential for success*. Even with a well-developed sense of purpose, the student may avoid engaging significant processes of thought because of a belief of inadequacy in one's chances for "succeeding." I observe lots of students who never seem to engage or assert their own thinking about the mathematics they are encountering, and I conjecture it is because they don't believe they can do so without exposing erroneous thought and thereby be seen as wrong and failing. Their activity in mathematics is self-constrained to imitative rather than generative levels because imitation involves the least amount of risk of error and perceived failure. Of course, from our perspective students might be succeeding at important thinking because they are aware of their errors, but the typical classroom climate for committing errors often seems negative. To improve upon this, both teachers and students must come to perceive the higher purposes for constructing mathematical knowledge and to view their roles in this constructive enterprise in ways that are very different from the current milieu.

Within any significant mathematical activity, one can experience both positive and negative emotions. It is important to acknowledge these feelings and to learn to manage and control their impacts. It is normal to experience sensations of frustration or defeat when attempting to solve a difficult mathematics problem, or feelings of confusion or being "lost" when confronting a complex, new mathematical development. Learning to cope with these debilitating feelings is crucial to persisting along productive lines of activity. The state of being "stuck" with its accompanying sense of panic must become an aspect of the investigative attitude, a stage of thinking which is manageable and which leads to potentially helpful actions. Thus, in the domain of emotional states a third element involves the willingness and capacity to *monitor and control the effects of feelings* (especially those that are negative).

Enquiry States

In this second domain of internal conditions which affect the quality of a mathematical experience are those aspects which involve the cognitive or intellective factors. In considering the benefits of any new experience, we would rely upon our *existing knowledge of subject matter*. Therefore, I would include the building up of meaningful, well understood ideas about number and space among the important factors influencing the conditions for quality experiences. This matter presumes to occupy the central purpose of the curriculum, though I have argued that the typical child's mathematical "tapestry" appears colorless, weak, and chaotic, woven of flimsy "threads" representing ineffective processes. Somehow, the major push to "learn content" is not helping students to become generative thinkers who can understand, remember, reconstruct, explain, apply, question, extend, and enjoy their mathematics.

To me, the central conditional element affecting quality experiences which could lead to growth as a generative thinker involves the *sense of being problematic*. Today the leadership in mathematical education has sought the adoption of problem solving as the central goal of the curriculum. But, what I am seeking here would hopefully go much deeper into the psyche of teachers and students. Thompson (1982) has reasserted what has been said by many others: "... to learn mathematics is to learn mathematical problem solving" (p. 190). While this might be misinterpreted as limiting problem solving to whatever one learns, it should imply the absolute centrality of a problematic perspective with each mathematical encounter.

This problematic perspective needs to pervade the sense of being in the school mathematics classroom. There is no problem "out there," written on the board or stated by the teacher. The problem to be solved is "within me," whether it is the product to be found, the triangle to be modeled, the procedure to be traced, the pattern to be discovered, the theorem to be interpreted, the generalization to be formulated, or the proof to be reasoned. The problem is essentially my internal condition. As I (the student) become aware of it, accept it, act upon it, struggle with it, nurture it, persist, make progress, perhaps solve it, and reflect upon it, I must be clear that the problem *is* integral to me. In our attempts to protect our egos or the teachers' attempts to protect the students (cf. Byers, 1984), we might chose to avoid such intimate, direct confrontations.

Through my interactions (i.e., both interiorized thought and external manipulations) my mental condition is in flux and change. In this sense, the problem is also in flux, undergoing transformation and alteration. It becomes me, and in some sense, I become it. When I experience mathematical situations problematically, then problem solving becomes integral to my conception of what it means for me to "do mathematics." My thinking is naturally enhanced. The cognitive processes we educators seek may occur more spontaneously. When I do not routinely experience mathematics problematically (even the textbook situations that are so labeled), I probably fail to develop the abilities or attitudes for engaging in problem solving, per se. It is so commonplace for students to resist and resent the textbook sections on "word problems" as departures from the more tolerable day-to-day activities.

A third conditional aspect of the enquiry states of a student involves the capacity and disposition to engage in self-conscious reflectiveness (Bruner, 1973). This aspect of mathematical experience has begun to receive greater theoretical attention (Thompson, 1985; Mason, 1982), but it appears to still be largely absent from observed classroom practices. Davis and Hersh (1981) describe how Lakatos (1976) in his classic *Proofs and Refutations* builds upon one of Polya's (1957) classroom dialogues to illustrate vividly his epistemological analysis applied to "informal mathematics," mathematics in process of growth and discovery. When students are engaged in intentional efforts to think about their experiences in order to highlight important actions and consequences, it seems that these perceptions have more potential for use on future occasions. This important mental activity can include attention to meta-knowledge issues for the student such as: How do I learn? How do I know when something is correct or "true"? What is the nature of my knowledge? What might I do to help myself learn more effectively? What effects do my feelings have upon my mathematical thinking? Are there ways of thinking which I seem to use? Can I become more intentional in my use of these modes of thought? Of course, students will probably need specific stimulation and encouragement to address such matters. Papert (1980) suggests that, given an environment which is rich in situations involving mathematical and scientific phenomena that can be explored playfully and constructively by children, they will naturally encounter points in their activity when they will consider epistemological questions.

Experiential Aspects of Constructive Computing Activities

My belief is that the students' constructions of mathematics can, indeed *must*, be enhanced and deeply influenced by the use of computers as learning and teaching tools. Effective applications of computing in school mathematics classrooms can positively impact upon the enquiry and emotional states of the students and the teacher. When students and teachers are provided with significant opportunities to explore mathematical concepts, procedures, and problems with the computer, it is not difficult to perceive many enhancements of the experiences they appear to be having (Hatfield, 1982). With powerful but simple programming languages, students can learn to construct their own computer procedures for modeling a concept, operationalizing an algorithm, or solving a problem. With carefully designed "microworlds" (Papert, 1980), the typically sparse classroom environment can become a stimulating, provocative mathematical world in which students can engage in experimentation and exploration of the conceptual schemes being modeled in the software (e.g., Thompson, 1985).

To suggest the possible enhancements of mathematical experience and the implications for teaching practice, curriculum reform, and research which instructional computing can raise for school mathematics, I will limit my discussion in this paper to two particular types of usages: student programming and simulating.

Student Programming and Algorithmics

Students of all ages are able to write and execute their own computer programs. It has been possible to find instances of precollege computer programmers for over twenty-five years. Today, with increasingly simple yet powerful computer languages and operating systems which are increasingly accessible on inexpensive microcomputers, even very young children are learning to compose their own computer programs (Papert, 1980).

Some mathematics educators have recognized the potential effects on the teaching and learning of school mathematics where student programming of the computer is an integral activity. In general, however, little change in either the subject matter contents or the approaches to teaching and learning school mathematics has occurred during this period in which more powerful computing has become potentially available to teachers and students.

We must now seriously address the issues and prospects for integrating student programming into mathematics studies at all levels of schooling. Further, the major "hidden" theme of school mathematics, namely "*algorithmics*," must be recognized and recast as a conscious, problematic emphasis in our approach to most topics.

To me, *algorithmics provides a key framework for thinking about the curricular rationales for integrating student programming activities into school mathematics.* "Algorithmics" is the study of algorithms and their construction. To most mathematics educators, this will appear simple and obvious. Surely, there exists a major attention in school mathematics to algorithms. Yet, there appear to be markedly differing notions of what might be involved in algorithmic mathematics.

Maurer (1984) identifies two meanings which seem to be used: "traditional algorithmic mathematics refers to performing algorithms; contemporary refers to creating them and to the mind set of thinking in terms of algorithms for solving problems and developing theory" (p. 430). This contrast emphasizes the distinction of conceptual meaning and behavioral performance (which has characterized one aspect of the continuing struggles between our desired and our actual classroom practices). And it underscores the important aspect of studying not only a "finished" algorithm, but its development. To me, what is sought in algorithmics is the crucial phase of *constructing one's own procedure*.

Perhaps we need to think about what we mean by an "algorithm," per se. To many, an algorithm is the object, the sequence of recorded steps embodied in a prescription or an example (Suydam, 1974). Teachers and textbooks refer to the written trace produced when one performs a procedure as "the algorithm." These days, we hear that a computer program is an algorithm. While from a realist view, this may make sense, we need to recognize that such references are communicating an existence for the algorithm apart from cognition (i.e., knowing and coming to know). And, in doing so, we are setting the stage for an adoption of non-constructive beliefs. In this realist view algorithms become interpreted as behavioral dictates to be attended to, observed, imitated, and performed.

To me, the psychologically significant qualities of an algorithm, which may help to indicate what a student actually knows, must be found largely in the person's activity of constructing the algorithm. When we inquire about the student's knowledge at a stage when his or her internalized method has achieved the refinements of an accomplished performance, we inevitably miss much of the richness of that person's experiences in arriving at that state. Testing or interviewing at this point probably fails to uncover:

1. the student's perception of the original algorithmic task which stimulates or "drives" the initial need for a method;
2. the exploratory paths followed and discarded;
3. the key decisions formulated, enacted, evaluated and re-formed;
4. the personalized reasons for performing a step;
5. the errors or mistakes that occur and the means by which the student recovers from them;
6. the basis, process, and nature of any generality of the method;
7. the perception of the class of tasks to which the procedure in formation may be applicable, including an awareness of special or limited cases;
8. the awareness of the structural "place" of the algorithm within the larger fabric of the person's mathematical tapestry; and
9. the disposition of the student toward algorithms and algorithmics as an aspect of the so-called discipline of mathematics.

In summary, algorithmics is a theme which is integral to constructing knowledge of mathematics. I believe that the need and effort to invent and understand algorithms is vital to the generative growth of one's mathematical ideas. Algorithmics and other more commonly recognized process-oriented, conceptual aspects can interact symbiotically in the dynamics of the student's mathematics---such attention to algorithms can foster the need for, and development of, new concepts and vice versa. "Delivering finished procedures" to the student may impede the achievement of such educational goals and effects. The context of formulating, testing, refining, and extending computer programs in response to viable algorithmic tasks can be a powerful approach to promoting the best features of an algorithmics theme.

Literally all subject areas include attention to procedures: "knowing how to." In most instruction, a single, inflexible, refined procedure is studied for an algorithmic situation. That is, only one "right" way to do it is presented; the method is portrayed as requiring a fixed sequence of dictated actions, and the approach is presumed to be the "best" possible, often being the culmination of other ignored developments. Typically, the steps of the procedure are exemplified or described by the textbook, and these are often modeled and explained by the teacher through examples. Student learning of these procedures is often expected to be imitative behavior.

Morley (1981) observes that "every branch of mathematics throws up its characteristic algorithms ... yet in school we seem to have lost any vision of the magnitude of the achievement of their construction, and so fail to convey it to pupils. Why is the challenge to *construct* the algorithms in the course of solving problems so rarely put in the classroom, before the showing and explaining begin?" (p. 50).

At the risk of overstatement, I will characterize what I perceive to be the dominant beliefs about the study of algorithms in school mathematics. First, educators' beliefs about the role of algorithmic knowledge in school mathematics are reflections of their attitudes toward their own experiences with algorithms. Among the more mathematically mature (such as secondary mathematics teachers or mathematicians), attitudes toward algorithmic knowledge can range toward cool and negative. The learning or performance of an algorithm is devalued, often appearing to be a "necessary evil." Some of these feelings are unconscious reactions, but much of the common reaction to algorithms is public and intentional. In either case, mathematicians and mathematics teachers can project these feelings in their communications to students, thereby nurturing similar attitudes.

We can speculate about the reasons for these attitudes:

1. Once one knows an algorithm, performing it doesn't seem to require much thought, and it is therefore devalued. If one is proficient with an algorithm, re-enacting the constituent skills feels trivial compared to other matters, such as analyzing the situation in which we are applying the algorithm. Habituated methods are seen to be simple retracings of well-worn behaviorial threads in the mathematical tapestry.
2. To many, the actual performance of a known algorithm is a burden, requiring monitoring of uninteresting details. "Grinding out" a result, once it may be clear what needs to be done, is often resented or trivialized, as it seems to divert one from the more valued problem or task at hand.
3. Performance of a familiar algorithm often seems to require, and thereby result in, lessened cognitive intensity and lowered attention to task. This may lead one to committing "stupid" mistakes, to which we often react with anger and irritation. These emotions are heightened when we cannot "find" our procedural mistake and we seem to "spin our wheels" on the larger, more worthwhile task.
4. Algorithms are often not considered to be "real" or significant mathematics (e.g., Halmos, 1985). That is, unlike other theoretical elements of an axiom system, algorithms are typically used only as a means to obtain understandings or results of the "important" end objects (axioms, definitions, theorems, proofs, and other solutions). The schism between "applied" and "pure" mathematics is partly a difference of attitude toward the value of an algorithm.

These conjectures about attitudes focus upon algorithms-well-known and do not address the aspect of algorithms-to-be-constructed (invented, discovered, proved). In contrast, we would conjecture that problematic approaches to algorithmics would feel as stimulating and valued as any good mathematics problem.

Second, because the primary recognized or valued "end result" of learning an algorithm appears to be recall and proficient execution, the dominant emphasis in the learning of an algorithm is imitation and memorization. Some attitudinal consequences for students of such learning of algorithms might be speculated:

1. Approaching the learning of an algorithm may feel like a mindless, pedantic activity resulting in doubts of its worth. This may be especially true when little instructional concern is given to motivating the need for a procedure, to developing rationales or understandings of the constituent steps of the method, to examining the range of applicable examples, to clarifying the special cases, to comparing and contrasting it to other possible methods, or to relating the use of the algorithm to interesting problematic situations.
2. When presumably learning an algorithm, students may grow weary of isolated drill and repeated practice on algorithmic exercises. This becomes acute when the intellectual gains feel minute and the purpose only seems to be to demonstrate "perfect" or "adequate" proficiency.
3. The connections that could be made between mathematical concepts or generalizations and algorithms in which they are applied or motivated are absent. This furthers the impression that algorithmic thinking is subservient or unrelated to "real" mathematics. It also weakens the mathematical powers of the student for anticipating the need for an algorithm or the applicability of procedures already known.

These speculations reflect feelings which may arise from any learning situation that lacks an appropriate degree of challenge for the participants to reason, to experiment, to conjecture, to understand, and to believe.

What about characteristics of student programming emphases in school mathematics? What viewpoints and practices are typical of mathematics instruction which involves computer programming?

1. The dominant curricular rationale in the U.S. seems to be that it is essential to offer computer programming in a comprehensive high school and that the place to do so is the mathematics department where a teacher can be found who has some programming skill. This view often results in an eclectic pot pourri of mathematical topics chosen to serve the cause of learning a computer language and certain elementary programming strategies and conventions.

More recently, we have witnessed attempts to shroud this questionable offering under a more respectable cloak of precollege "computer science."

2. Of lesser occurrence is the occasional inclusion in the textbook of a computer program, typically listed and discussed as a "finished" procedure. While these usually do relate to the concept or procedure under study at the time, it is common for students to have no "hands-on" experiences with the programming situation (and in many classes the program may never even be demonstrated to the class with an actual computer!). These encounters seem to provide the student with an initial acquaintance with the symbolism of programming but may do little to further the more significant knowledges of algorithmics.
3. In only isolated instances have student programming activities been integrated into the scope and sequence of the mathematics curriculum (e.g., Feng, 1973; Johnson, et al. 1968). In such treatments the balance of treating the mathematical ideas and the programming syntax remains a major issue. Also, there has been little alteration in the curricular contents as a consequence of student programming, with the traditional topics maintained but treated with an emphasis upon "teaching the computer." Thus, even in places where teachers are attempting to involve students in programming activities in behalf of mathematical learning, the curricular topics have not been chosen to reflect the potential advantages of computing. There develops another "curricular schizophrenia" in which teachers try to serve competing dictates for goals, pedagogy, and accountability.
4. Most mathematics teachers have had little or no experience in their own mathematical studies where the production of a computer program was an activity in pursuit of their own understanding of a concept or solving of a problem. And most professionalized teacher education courses do not specifically address the pedagogical aspects of teaching school mathematics through a student programming emphasis. As a result, most mathematics teachers are at a loss about how to conduct an effective mathematics lesson involving student programming.

Successful mathematics students (and teachers) do, indeed, study and "learn" a fairly large number and variety of algorithms in order to continue to achieve in their mathematical studies (and in their teaching). Yet, we should ask: What is the quality of their knowledge? What are the aims and purposes of the learning (and teaching) of these algorithms? What experiential qualities characterize the learning (and teaching) of these algorithms? And what are the attitudes and dispositions of the learners (and the teachers) toward this required or acquired knowledge?

On the other hand, many of our students have difficulties learning or applying the algorithms we expect them to know, and the poorer quality of their algorithmic knowledge is often cited as a barrier to further growth of mathematical ideas. Again, we should ask: What is the nature of this learning (and teaching) which might explain poorer conceptual or performance results? What is the nature of the intellectual requirements to know algorithms in order to construct further mathematical knowledge?

To most thoughtful mathematics teachers, it is *the quality of the student's mathematical knowledge* that is always the crucial matter. And, to me, a crucial factor determining the quality of knowing is the quality of the students' experiences in constructing their knowledge. Too often we consider only aspects of knowledge that focus attention upon the concept identified or exemplified, the generalization recalled, the problem solved, the theorem proved, or the procedure executed. There, the emphasis is upon the mathematical "object" or the result, rather than upon the processes of construction and reconstruction of the idea. Yet most mathematics teachers realize that it is far more important to be able to rebuild (i.e., construct or derive again) the forgotten or confused result than to rely only upon simple memory of previously studied items of knowledge. Indeed, ability to solve mathematical problems requires such competence to construct a "method" (i.e., a solution path along which one has never quite passed before). Wittrock (1973) emphasized that learning with understanding is a generative process, and Piaget (1973) asserted that to understand is to be able to invent.

To me, an essential aspect of algorithmics in school mathematics is its problematic, constructive quality. The learning of an algorithm should be a problem-solving activity. The effective teaching of mathematical algorithms should incorporate problematic, constructive approaches. Osborne (1974) argues that in a world characterized by extensive use of the computer, individuals must understand algorithmic processes, and mathematics educators must nurture the development of intellectual paradigms which avoid limitations on the mathematical imagination and creativity of the individual. He asserts that present school mathematics programs, along with the associated research, do not address the issue of goals and activities fitting children's future adult needs in mathematics and that this is especially the case with algorithmic thinking.

Maurer (1983, 1985) presents strong arguments of "an algorithmic way of life," and the needed debate in the mathematics community has only begun (e.g., Douglas, 1985; Hilton, 1985; Renz, 1985; Fey, 1984). He, too, contends that an algorithmic theme should be central to curriculum revisions and suggests that the following steps be taken:

1. describe traditional mathematics in algorithmic terms;
2. discuss the correctness and efficiency of algorithms;
3. present algorithms as a method of proof; and
4. present modern ideas of algorithms in which precision of expression and recursive techniques are embodied.

He urges that "thinking with and about algorithms can unify all one's mathematical endeavors. It also extends the range of one's mathematics and provides as satisfying an esthetic as the old existential esthetic--if one will only let it" (1985, p. 2).

Developing a knowledge of algorithmics (as the study of algorithms and their construction) should involve at least the following experiences:

1. understanding that algorithms are essentially generalized procedures of an individual's method for solving a class of problems;
2. realizing that for an algorithm to be deeply meaningful, one must understand its origin and construction;
3. accepting the need to formulate and refine algorithms as a way of life (Maurer, 1985) integral to the developments of new mathematical ideas and their applications;
4. believing that algorithms, as idiosyncratic methods, are necessarily constructible by each person who studies mathematics;
5. participating in the social construction of "shared" algorithms as a valuable context for clarifying and refining one's own personalized methods;
6. understanding that while the construction of one's personalized algorithm for a particular mathematical task is a worthwhile activity, it can also be important to transform one's method to more conventional forms; and
7. reflecting upon one's experiences with a particular algorithmic construction (or series of constructions) to consolidate understandings of the individual algorithms and their possible interconnections, to abstract significant generalizable structural qualities of algorithms, and to ponder important intellectual processes involved in building and refining algorithms.

Teaching school mathematics in which integrated student programming activities serve goals of constructing knowledge of algorithmics should involve at least the following processes:

1. identifying an appropriate algorithmic task with strong connections to the conceptual aspects of the mathematical contents under study;

2. providing the possible experiences which establish the conceptual anchors for any algorithms to be constructed, including understandings of the preliminary concepts, of the applied or theoretical situations in which a procedure is needed, and of the anticipatory "action schemes" which may be required to approach the development of an algorithm;
3. setting the algorithmic task in a problematic fashion where the student understands and accepts the challenge as worthwhile and solvable;
4. guiding the student's efforts within a constructive methodology in which conceptual challenge and stimulation are tempered with support and respect of idiosyncratic approaches;
5. helping students to accept their own initial production of a computer program for some task (however crude, limited, or "bugged") as an "active object" on which they can test and revise their ideas;
6. helping students to develop dispositions and tactics for identifying and recovering from the errors and limitations of their computer programs;
7. provoking students to seek more generalized or elegant formulations of their computer algorithm; and
8. guiding students to make reflective abstractions about related algorithms and about their intellectual processes used in solving algorithmic problems.

The research implications and needs reflected in these advocacies are manifold and complex, and I will not address them in the detail they deserve. Yet it must now be an important priority for us to establish research agendas to include attention to the issues and questions involved in the teaching and learning of school mathematics with an emphasis upon student programming and algorithmics. Among the types of investigations needed, I would encourage our discussion of the following questions:

1. How do students understand mathematical variables and expressions as they are operationalized in computer algorithms? What conceptual conflicts may arise with computing variables and regular algebraic analyses?
2. How are student's understandings of traditional concepts, such as fraction, area, congruence, limit, continuity, or proof, affected by algorithmic experiences? In particular, what advantages and disadvantages occur when dealing with the infinite in mathematics and the finite machine?

3. How do students think about algorithmic concepts, such as iteration and recursion, and what effects does such knowledge have upon common mathematical situations, such as solving equations?
4. How do students feel about approaching their studies of mathematics more constructively where algorithmic tasks are featured as a context for building and refining one's own procedures as embodiments of personal concepts and theories?
5. What metacognitive demands and effects may arise within student programming and algorithmic emphases (e.g., Papert, 1980)?
6. What are salient qualities of the actions of teachers who integrate student programming and algorithmic emphases into their lessons? How do these teachers balance the need to teach programming with the essential goals of mathematical education (e.g., Hatfield, 1985)?
7. What pedagogical beliefs and personal theories allow teachers to open up to modernization of their curriculum?
8. How might the curriculum contents be affected by the greater power of computing available to a student who understands how to use the programmable machine to explore mathematics?

Research is needed related to student programming and algorithmics given the tremendous potentials of enhancing teaching and learning with microcomputers. There continues to be considerable skepticism about the worth of mathematics students writing and using their own computer programs. Most importantly, there is the fundamental need to confront a reconstruction of our goals and meanings of educating in mathematics. In pursuing research to explore algorithmics and student programming, we must openly construct and study alternative pedagogical and curricular paradigms.

Simulating

Computer-based simulations[2] for teaching and learning mathematics represent a powerful application of today's technology. Such programs are designed to represent abstracted features and functions of some complex, challenging, dangerous, expensive, time-consuming, or puzzling phenomenon. By executing such programs interactively, students and teachers can experience significant constructive approaches to building-up mathematical ideas. These experiences can involve the enhancement of the students' emotional and enquiry states. Most of the time, these programs are selected and acquired ready-to-use in the classroom. However, I will also include situations in which the goal is to design and encode our own programs for simulating; this will connect student programming and simulating as powerful mathematical activities.

With prepared simulation programs, the focus will be upon executing the program in order to study the circumstances and events of the phenomenon as it is being modeled. Among the features of an effective simulation program, the following can be cited:

1. The behavior of the executed program represents an accurate, high fidelity (though probably simplified) portrayal of the "real" (non-computer) events as they are understood by experts.
2. The underlying model incorporates important features and variables which function in ways that imitate such conceptual components to be "found" in the "real" phenomenon.
3. In some cases the manipulation of the model will involve gaining control over sources of "noise" or interference from factors which may confound or disrupt the situation.
4. To permit manipulation and exploration, important variables or parameters in the model can be altered by the user.
5. Useful records of inputs and outputs, as well as interrogation of intermediate conditions and manipulations, may be provided.

Important qualities of a constructive approach to learning can be incorporated into the use of simulation programs in the mathematics classroom. However, this will surely depend upon *how* such activities occur. To me, this is the critical matter for all instructional computing activities. Though some have advocated computer-based instruction as a replacement for the teacher, I believe that the social dynamic of a knowledgeable, sensitive adult who is committed to aiding students in developing their knowledge is necessary to achieve quality mathematical experiences. The use of the computer should be an enhancement of the essential mathematical dialogue in which the power of the machine serves to extend our thinking.

Thus, the use of a simulation program must be within the ongoing context of mathematical ideas under exploration and construction. In this sense the students should have an experiential framework in which an episode with the simulation program will fit. They would be armed with important background concepts and questions which might then be enriched through the simulation. Students should have considerable opportunity to make suggestions for entering values which influence the operation and outcomes of the simulation. As argued, they should be encouraged to become intimately connected with the problem of understanding, for that problem is ultimately theirs.

By shifting the focus to designing and building of simulators, we can offer situations that seem to be of considerable interest to students[3]. Within a curriculum where students have learned to produce their own computer programs, we can find many mathematical situations which can be posed as problems in designing, encoding, and using one's own simulation program.

For example, at the time middle school students are studying the idea of area of a region, we typically restrict the discussion to simple polygons and emphasize the traditional Euclidean formulas. But I like to encourage students to consider the means by which we might find the area of any shape. They will suggest interesting tactics for subdividing, measuring, and counting which often appear to involve informal precursors to sophisticated notions of sequences and limits. Using a dart board analogy, they can quickly discover the essential features of a Monte Carlo method. Sometimes we will set up paraphernalia for experimenting with the method, but we soon realize that the conduct of the many trials needed to get a good estimate is laborious. When they have had previous experiences with the (pseudo) random number generators of the microcomputer, it becomes a relatively direct matter to design and encode an algorithm for simulating a large number of "darts randomly thrown." With our program we can then experiment with test cases (with known areas) to establish how many darts are needed to obtain a satisfactory estimate of the area. Once we have a sense of confidence in our method, we can then apply it to a wide range of provocative and sophisticated examples (e.g., confirming an estimate of pi, finding the area under a curve graphed from a complicated function, or seeing how cartographers might use the technique with satellite photographs to determine land areas).

In using such situations with many students, I cannot pretend to know any of the experiential consequences. But I can speculate about a few from my recalled inferences during these situations. The impacts upon their concepts of area have been worthwhile, for the students seem to understand and appreciate the fundamental measurement task from another (probabilistic) view. The sense of involvement with the mathematics of area and chance is probably greater than when I simply demonstrate such a simulator to them. They often surprise me with the questions and suggestions they make for estimating such areas, for building the simulator, and for the subsequent uses they want to make of it. When encouraged to do so, some students will attempt to refine the program to make it more interactive, efficient, or general (e.g., 3-dimensional). I am sure that my impressions of their experiences are biased by my own experiences in these situations.

Summary

My ongoing search for clearer understanding of what students do, or might, experience in their constructions of their mathematics is inevitably framed in terms of my own experiences as a student and a teacher of mathematics. My advocacies of what students "should" experience in building-up mathematics are primarily determined by my idealized conceptions of the consequences these experiences might have upon potentially significant intellective and emotive processes.

I have expressed my discouragement over the general lack of *quality* in the mathematical situations students seem to experience in today's classrooms. To change this, we must focus upon the total "mathematical life" of the student, the genesis, development and nurturing of his or her "tapestry" of mathematics. While external factors can surely matter, the major determinant of quality in mathematical experience involves the internal conditions which reflect the emotional states and enquiry states of the student. I have noted only a few of the possible aspects of these conditional influences.

The effective application of the computer represents one possible external factor in our attempts as teachers to enhance school mathematical experiences. Though fully cognizant of the multiplicity of usages of the computer with mathematics students, I have chosen to limit this discussion to student programming and simulating activities. My vision of the role and purpose of student programming is anchored in my conceptualization of algorithmics as a major theme in the curriculum. My ideas for incorporating computer-based simulations include attention to student-generated simulators. Both types of computing permit, and flourish under, a problematic approach which is essential to enacting a constructivist pedagogy.

The consequences of lesser quality experiences in school mathematics are to be found all around (and perhaps within) us. It is highly significant that scholars and practitioners concerned with educating for mathematics are now showing concern for understanding mathematical experience---ours and our students'.

Notes

1. I am reminded of the fascinating struggle with the idea of *quality* which the complex character Phaedrus experiences in Robert Pirsig's novel *Zen and the Art of Motorcycle Maintenance*. I won't attempt to justify my own focus upon quality in this discussion but will acknowledge and leave unresolved the inevitable philosophical dilemmas resulting from attempts at defining this construct.
2. One of the complicated aspects of dealing with computer-based simulations as an instructional approach involves the question of what types of programs or systems are to be included. Essentially, every program is a simulation of something which could be found in other forms (though the speed, precision, or scope may be very difficult to duplicate without the machine). Indeed, it is possible to

think of the programming language and operating system as a simulation (e.g., the Turtle Geometry features of the Logo language were originally designed to control a floor device but now produce graphical representations of the "drawing" actions). While I will not attempt to define the notion of a computer simulation precisely, the general intention to model some phenomenon in order to learn more about its features and functions will have to suffice as a guideline.

3. One can find an increasing number of software "tool kits" for making up such simulations. These systems typically offer the student graphical icons for items such as gears, wheels, cubes, or other solids; triangles or other polygons; arrows, pulleys, springs, balances, number lines, or tree diagrams; or icons denoting processing for performing operations upon the items, such as fitting together, taking apart, turning, factoring, reflecting, bisecting, making perpendicular, or simplifying. These primitive icons can be selected, assembled, and manipulated as situations for exploration. Such "microworlds" have been widely attributed to the ideas of Papert (1980) and his associates.

12 To Experience is to Conceptualize: A Discussion of Epistemology and Mathematical Experience[1]

Patrick W. Thompson

When I accepted the invitation to comment on the papers in this volume, I had little idea of their diversity. Yet, that very same diversity, while being initially overwhelming, turns out to be a considerable strength of the collection. It is remarkable that these papers, written within diverse traditions and disciplines, reflect a coherent theme: experience and conceptualization are inseparable.

In assimilating sense data or accommodating to it, we cannot experience "the world" without already "knowing" something about it. This is not to say that what one knows is correct, true, or even viable. Rather, it says only that we must already know something with which sensation or conception resonates.

In the same vein, we cannot experience the world mathematically without using mental operations we would call mathematical. Let me anticipate two interpretations of this statement. The first is that the mathematics anyone comes to know is innate, ready to emerge over time, awaiting appropriate environmental "triggers." The second is that no one may know or come to know mathematically, which is evidently absurd. Neither interpretation is consistent with a constructivist epistemology, yet the statement that led naturally to them *is* a hallmark of constructivism.

Thus, there are two principal challenges implied by a constructivist epistemology of mathematics. The first is to provide compelling arguments that it is *possible* for adult mathematics to emerge as the product of life-long constructions, where those arguments are painstakingly, evidently, non-Chomskian. The second is to hypothesize *constructive mechanisms* by which specific knowledge might be made and to give detailed accounts of those constructions. The papers by Bickhard (chap. 2, chap. 5), Cooper (chap. 7), Steffe (chap. 3), and von Glasersfeld (chap. 4) address these challenges with remarkable clarity.

A third challenge to a constructivist epistemology of mathematics is more practical, and at the same time, it is the more important challenge. This is the challenge of framing curriculum and pedagogy within a constructivist tradition. It sounds quite non-constructivist to say that, as mathematics educators, what we try to do is shape students' mathematical experiences. Yet, that is what mathematics educators working within a constructivist framework try to do. We attempt to provide occasions where students' experiences will be propitious for expanding and generalizing their mathematical knowledge. Not just any experience is satisfactory.

Five papers take up the second challenge. Steffe (chap. 3) and Dubinsky (chap. 9) address models of students' knowledge in specific mathematical domains. Hatfield (chap. 11) addresses pedagogy. Confrey (chap. 8) and Cooper (chap. 7) address both. Each paper informs our attempts to characterize curriculum and pedagogy within a constructivist tradition.

The paper by Kieren and Pirie (chap. 6) can be profitably viewed as belonging to a separate, important category--methodology. I do not mean methodology in the limited sense normally used in experimental psychological research. Rather, I mean it in the sense of what is needed to give sensible accounts of observations. Kieren and Pirie (chap. 6) are concerned with methods of explanation, and at the same time, they use their language of explanation (recursion) to describe desirable experiences had and to be had by students.

Dubinsky (chap. 9) raises issues quite relevant to this notion of methodology, but his paper is not an analysis of method. Instead, it is a dialectic. He analyzes the processes of delimiting a class of phenomena that need explanation while constructing a framework for describing them.

Lewin's (chap. 10) paper was the most difficult for me. It provides a clear demonstration of methodology in that he uses his theoretical foundations to explain the sense-making activities of students' readings of literary text. However, it is more profitably viewed as a challenge to mathematics educators to cast mathematics education as *paideia*, as being fundamentally concerned with the formation of character. In later remarks I will suggest that this challenge is entirely consistent with mathematics educators' concern with the provision of occasions for students to have rich, meaningful mathematical experiences.

Construction of Mathematical Thought

Constructivism is commonly thought of as an epistemology--a theory of knowledge. Constructivism has another face--it is a theory of the genesis of knowledge. It is emerging as a theory of learning. It makes specific the claim that anyone's knowledge is the life-long product of constructions.

As a learning theory, however, constructivism is in its infancy. It is seen by many as being more useful as an orienting framework than as an explanatory framework when investigating questions of learning. To say that we are constructivists only because we believe that knowledge is constructed and not received is less than compelling, and it is clearly not useful. It is also insufficient to argue that a psychological theory is invalid if it presupposes direct access to "reality." This argument has been around since the Skeptics. What we need is a *technical* constructivism. We need a technical constructivism that allows its proponents to form precise, testable hypotheses and that allows its opponents the opportunity to refute them, and to refute them on the basis of the adequacy and viability of the *system* of explanations constituting constructivism.

Taken as a collection, the five papers by Bickhard (chap. 2, chap. 5), Steffe (chap. 3), von Glasersfeld (chap. 4) and Cooper (chap. 7) constitute a primer in constructivist learning theory. Bickhard (chap. 2, chap. 5) lays a theoretical foundation for a constructivist learning theory, and Steffe (chap. 3) proposes mechanisms for the construction of arithmetical knowledge. Reflective abstraction, which is central to both papers, is clearly explicated by von Glasersfeld (chap. 4), as are several other explanatory constructs. Cooper (chap. 7) investigates the roles of repetition and practice as constructive mechanisms in learning. These papers are fundamental reading for any student of mathematical learning and cognition.

Foundations of a Constructivist Learning Theory

Bickhard (chap. 2, chap. 5) and Steffe (chap. 3) let us glimpse a technical constructivism. Their papers attack the problem of the *possibility* of learning in general and of learning mathematics in particular. That the possibility of learning is problematic can be appreciated upon consideration of how an individual might construct knowledge that is not made by associations of existing concepts. Both Bickhard (chap. 2) and Steffe (chap. 3) respond to Fodor's anti-constructivist argument: if learning must involve the construction of new representations, then learning cannot happen and "some basic set of representations, combinatorically adequate to all possible human cognitions, must be innately present" (Bickhard, chap. 2, p. 15).

Bickhard (chap. 2) claims that Fodor's argument is valid only if one accepts encodingism--the idea that humans somehow have mental symbol systems that are isomorphic to features of reality. He argues that encodingism is, in fact, an incoherent position, and that, in this regard, Fodor's argument for innatism is fallacious.[2]

Steffe takes a different approach to refuting Fodor's innatist conclusion. He attempts to offer a counter-example to Fodor's argument. It is debatable whether Steffe's (chap. 3) example is a counter-example to Fodor's argument or is instead a demonstration of an alternative framework having its roots in constructivism. But this is a minor point. What is clear is that Steffe (chap. 3) offers an interpretation of one student's learning that is not subject to Fodor's argument.

Both approaches succeed by denying naive realism--the idea that somehow we are imprinted with knowledge of the real world--at the foundation of their theories of representation. Instead of characterizing representations as representing something about the real world, Bickhard (chap. 2) and Steffe (chap. 3) characterize them as "mental stand-ins," *made by the individual doing the representing*, for interactions with a world of objects or ideas. That is, we can represent by re-presenting. These ideas are new and old. They are new because of context and specificity. They are old in the sense that Piaget (1968) anticipated the need to address issues of representation to make specific his claim that language is just one expression of "the general semiotic function" (Piaget, 1950).

Piaget delineated three forms of representation: indices, signs, and symbols. An *index* is a re-presentation of an experience by recreating parts of it in the absence of the actual experience (e.g., imagining an exchange of hands in a "promenade round" to re-present a square dance, or rhythmically nodding one's head to re-present the experience of counting).

A *sign* is a figural substitute--something that captures an essential aspect of a class of experiences, but which is only analogous to them in its similarity. The perception of a wavy line on a yellow board alongside a highway signifies to many people that a bend in the road lies ahead. The wavy line in Figure 1 has nothing to do with one's experience with roads as such, yet it suggests a feature of one's experience of driving on winding roads. Similarly, the underline character in "2 + _ = 7" is not usually part of one's experience in

Figure 1: A road sign.

carrying out arithmetical operations, yet it suggests that something is missing.

Signs are inferentially linked with their referents, but the inference is much less direct than is the case with indices. I suspect that, were we to look closely, we would find that even the most sophisticated knowers of mathematics make abundant use of signs in organizing their mathematical knowledge.

A *symbol* represents something only by way of association. Symbols have the qualities of arbitrariness[3] and, in the case of symbols which serve a communicatory function, conventionality (Hockett, 1960; von Glasersfeld, 1977, chap. 4). It would be presumptuous of me to try to improve upon von Glasersfeld's (chap. 4) discussion of symbols.

While Bickhard (chap. 2) focuses primarily upon issues of representation, Steffe (chap. 3) focuses primarily upon issues of learning, in particular on accommodations that can account for learning. Before judging the success of Steffe's (chap. 3) attempt, we should remind ourselves of his principal goal--to establish that learning is *at least* inductive inference and to give an existence proof that it can be more. To judge it as successful, we need to answer two questions affirmatively: Do we accept that Steffe's (chap. 3) ideas of engendering and metamorphic accommodation as viable, explanatory constructs? Does metamorphic accommodation account for a change in Tyrone's behavior that inductive inference cannot? These are non-trivial questions. When answering them for yourself, you will come face-to-face with the core of Steffe's theory of units and operations.

One thing that seemed missing from Steffe's (chap. 3) analysis was specificity. This might sound like an odd comment, especially given the extremely small segments of behavior analyzed in great detail by him. But the kind of specificity I have in mind is different from what Steffe (chap. 3) gives us. I would like to have an image of Tyrone's *knowledge*. It is evident that Steffe (chap. 3) has such an image, but it is not well communicated by natural language. I believe we can take advantage of decades of research and methodology in artificial intelligence and information processing theory. Models expressed in natural language are notoriously poor at facilitating precise thought and communication. Also, they are extremely cumbersome when one is trying to capture the dynamics of functioning systems. I am reminded of Cobb's (1987) well-known remark that "it would be a tragedy if all serious students of cognition felt compelled to express their creativity solely within the confines of particular formalisms such as computer languages." It would be just as tragic if all serious students of cognition eschewed formalisms such as computer languages.

Reflection and Repetitive Experience

Reflective abstraction is an idea that is central to constructivism. Bickhard's (chap. 2) and Steffe's (chap. 3) arguments would have gotten nowhere without appealing to reflective abstraction. Without something like it, constructivist theories of learning are dead in the water. In constructivism, reflective abstraction is the motor of accommodation, and hence of learning.

To say that reflective abstraction is central to constructivism is one thing; to say what it *is* is quite another. At times, discussions of reflective abstraction take on the character of describing the homonculus--the little man in the mind that does all the nasty work not accounted for by a cognitive theory. We have been much more successful in describing mechanistic models of the products of reflection than we have in describing how people reflect.

All this notwithstanding, we need to have a clear idea of how any model of reflective abstraction needs to behave, and we need to have a clear idea of the phenomena we wish to ascribe to the operations of reflective abstraction.

Von Glasersfeld (chap. 4) gives us a portrait of reflective abstraction in these regards: its necessity in constructivism, its character, its history as developed in Piaget's genetic epistemology, and the similarities between reflective abstraction in Piaget's theory and in Locke's empiricism.

What von Glasersfeld (chap. 4) makes clear is that reflective abstraction, re-presentation, and representation are inseparable aspects of cognitive functioning. If I can add to von Glasersfeld's (chap. 4) contribution, it is this: Piaget gave considerable prominence in his earlier work to the development of intuition, and I believe it was for a good reason. By focusing on intuition, we gain additional clarification of the ideas of reflective abstraction and at the same time push the homonculus farther into the background.

As is frustratingly common with so many terms appearing in Piaget's writings, he failed to give a clear definition of what he meant by "intuition." The clearest statement I have found is in (Piaget, 1950).

> We see a gradual co-ordination of representative relations and thus a growing conceptualization, which leads the child from the [signific] or pre-conceptual phase to the beginnings of the operation. But the remarkable thing is that this intelligence, whose progress may be observed and is often rapid, still remains pre-logical even when it attains its maximum degree of adaptation; up to the time when this series of successive equilibrations culminates in the "grouping," it continues to supplement incomplete operations with a semi-symbolic form of thought, i.e. intuitive reasoning; and it controls judgments solely by means of intuitive "regulations," which are analogous on a representative level to perceptual adjustments on the sensori-motor plane. (p. 129)

To draw out the significance of intuition, I need to digress briefly. One modern interpretation of constructivism is in terms of autopoietic systems (Maturana & Varela, 1980; Maturana, 1978) and as cybernetic systems (MacKay, 1969; Powers, 1973, 1978; von Glasersfeld, 1976, 1978). Within these perspectives, cognition is viewed as the product of a nervous system's attempts to control and regulate its functioning. Of course, "its functioning" is not something that happens with no exogenous intrusions.

The primary aspect of autopoietec or cybernetic systems is the fundamental, overriding principle of control: the elimination of perturbations within the system, the resolution of unmet or unattainable goals. Intelligence progresses through the development of cooperative systems, or schemes, for eliminating classes of perturbations ("classes" from the cognizing organism's perspective). These schemes--systems for controlling cognition--while emerging, fit roughly with Piaget's description of intuitive thought.

Intuitive thought, then, is the formation of *un*-controlled schemes which themselves function to control aspects of cognitive functioning. But these un-controlled schemes are themselves part of the organism's cognitive functioning, and hence are something to be controlled. They become regulated as *their* controlling schemes reach the level of intuition, whence the schemes controlled by them become equilibrated. That is, intuitive thought is actually the fodder of operative thought. Figure 2 illustrates this discussion: the emergence of intuitive thought and then intuitive control of intuitive thought--operative thought.

Attention to intuition has two benefits. First, the homonculus is now extrinsic to our picture of the emergence of knowledge. The homonculus is in the principles of cybernetics. The question now is one of architecture--how are biological systems organized that they might or might not behave like this? Second, understanding intuition as unregulated schemes provides a connection between, on the one hand, von Glasersfeld's (chap. 4)

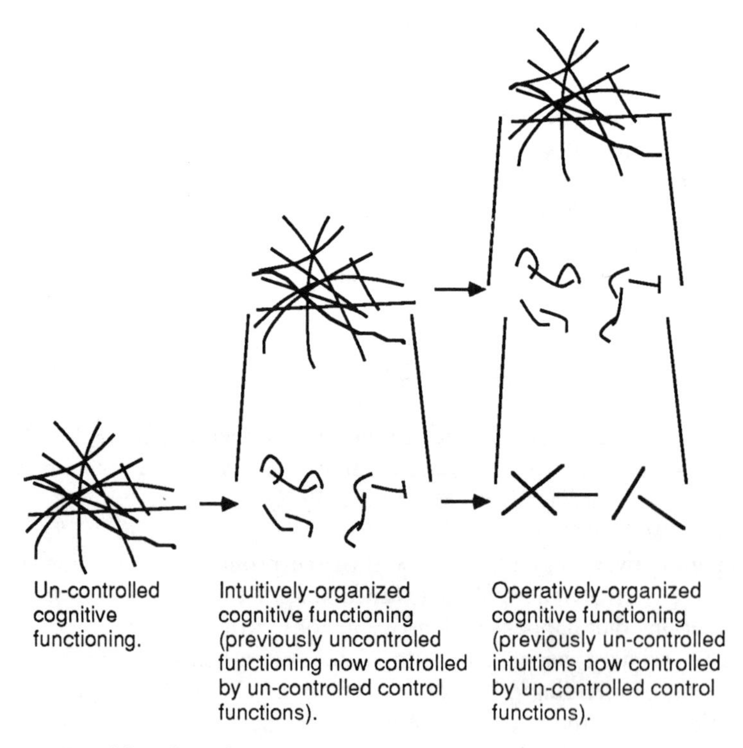

Figure 2: Intuitive thought.

characterizations of reflective abstraction and reflected thought and Steffe's (chap. 3) characterizations of metamorphic accommodation, and, on the other hand, Cooper's (chap. 7) descriptions of the dramatic influence of repetitive experience. Intuitive thought develops through recurring experience by way of functional accommodations; it is transformed into operative thought through reflective abstraction and metamorphic accommodation.

Curriculum and Pedagogy

"Constructivist curriculum" and "constructivist pedagogy" sound like oxymorons. It seems paradoxical that, on one hand, we maintain that we are, of necessity, in the dark about how and what people think--that people will make of their social and physical environs what they will, while on the other hand, we plan what students are to learn and attempt to design "effective" instruction.

The paradox is in appearance only. We put on the hat of constructivism *so that* we have more coherent visions of what *might* be happening when students evidently learn and understand mathematics and think mathematically. It is from a basis of coherent visions that we are positioned to have greater confidence in our plans.

Cooper (chap. 7), Hatfield (chap. 11), and Confrey (chap. 8) each inform our attempt to enrich our understanding of what it means to learn, understand, and teach mathematics. Cooper (chap. 7) sets out to convince us that appropriately-conceived repetitive experience can provide a strong foundation for reflective abstraction, and hence that it is a crucial element of students' mathematical learning.

Hatfield (chap. 11) reminds us that experience is a private affair and that there are many affective components to mathematical experience. Nevertheless, it seems that a common thread to his arguments is that "good" experiences lead students to feeling in control of situations or at least lead them to feel confident that they can come to be in control. His discussions of student programming and simulations explicate the powerful idea that one avenue toward building control over ideas is to operationalize them.

Confrey (chap. 8) makes concrete the adage that we cannot understand students' behavior without understanding at least one student. She attempted to understand the thinking of one student, named Dan, in the relatively complex domain of exponential functions. Her analyses of this one student's difficulties gives insight into the machinations of "correct" knowledge of exponentiation. To paraphrase Hersh (1986), it is not until we see students who provide counter-examples to our implicit assumptions about the constitution of specific concepts that we recognize them and make them explicit .

Practice Space and Mathematical Knowledge

Cooper (chap. 7) declines to use the word practice because "the term practice connotes that the activity being practiced is the skill of interest" (p. 105). This is a wonderful distinction. "What is being practiced" is normally in the mind of an adult observer. We *intend* something to be learned, and we have students "practice" it. What they actually learn can be quite a different matter.

We can still use the notion of practice in our theoretical and pedagogical analyses, however. Instead of beginning with the statement "Students practiced *X*," we should begin with the observation "We *intended* that student practice *X*," and then continue by asking the question, "What did they *actually* practice?" What they actually practiced is probably what they actually learned.

The examples given by Cooper (chap. 7) are compelling. They also help to clarify an important relationship between knowledge and reflected knowledge. The more tightly woven intuitive knowledge is, the richer is knowledge constructed as a reflection of it. The "map"--the set of interrelationships among situations to which Cooper (chap. 7) refers--is what is reflected. The denser the populated areas, the more relationships in the map. The fewer the populated areas, the sparser the map.

This raises an issue. If repetitive experience generates "richly interconnected spaces," providing a foundation for reflection and reflected knowledge, then as mathematics educators we have a responsibility to describe "spaces" we hope get constructed. This is a curricular issue, and one not settled by Cooper's (chap. 7) paper. Cooper (chap. 7) makes evident the need for rich and varied repetitive experience in students' schooling. However, to settle on the interconnections we wish children to generate through repetitive experience, we must clarify what we hope they achieve. We need to describe *cognitive objectives* of instruction, and we need heuristic guidelines for organizing instruction and curriculum so as to have some confidence that the objectives can be achieved. How shall we vary situations to map a space? How shall we decide whether two variations are within the same space?

Cooper's (chap. 7) analyses prompted me to recall comparisons of American, Japanese, Taiwanese, and Soviet mathematics textbooks (Fuson, Stigler, & Bartsch, 1988; Stigler, Fuson, Ham, & Kim, 1986). The gist of these comparisons was this: American textbooks rarely have word problems of any complexity, the problems are commonly of the lowest order of conceptual difficulty, and problems within a set hardly vary in their solution procedure (Porter, 1989). What do students practice when they "work" these problems? At best, they practice "getting answers." At worst, they practice ignoring such things as context, structure, and situation. In any case, students do not have occasions to generate the "richly interconnected spaces" that

Cooper (chap. 7) has identified as being crucial for constructing mathematical knowledge. They end up with islands of superficial knowledge without a canoe to get from one to another.

Algorithmics and Mathematical Knowledge

A common view of "skill" in mathematics is to know a large number of procedures for solving a similarly large number of problems. The word "algorithm," interpreted from this viewpoint, justifiably strikes fear among teachers and students. Here is a brief list of "school-math" algorithms:

1. whole-number addition
2. whole-number subtraction
3. whole-number multiplication
4. whole-number long division
5. whole-number short division;
6. all of the preceding with fractions instead of whole numbers;
7. all of the preceding with decimals instead of whole numbers.

Add to this list all the variations that school texts commonly promote: addition with and without trading, subtracting "across zero," division by a single-digit number, division by a two-digit number, and so on. We soon see a combinatorial explosion in the number of "algorithms" students meet. To ask anyone to learn such a large number of isolated, ostensibly unrelated procedures is inhumane.

We are urged to view algorithms from a different viewpoint, to consider that significant mathematical learning takes place when students' *create* algorithms and when they investigate them systematically. The task Hatfield (chap. 11) has taken on is difficult. He must communicate the richness contained in his idea of algorithmics when many in his audience have not experienced this approach.

Hatfield's (chap. 11) call for the inclusion of algorithmics--the creation and study of algorithms--in school mathematics might sound reminiscent of discovery learning (Bruner, 1963; Hendrix, 1961), but it is actually quite different. Discovery learning emphasized the "uncovering" of concepts and principles, as if they were to be found by turning over a rock. Algorithmics emphasizes the routinization of problem solving by the creation of schemes. At first glance, to routinize problem solving sounds self-contradictory, but it is a standard activity when one is *doing* mathematics. Schemes contain concepts and principles. To create schemes, one must also create component concepts and principles, but that in itself is insufficient. One must also create operations and relationships between and among them.

Hatfield (Chap. 11) also concentrates on student programming as a vehicle for bringing algorithmics into school mathematics. His efforts are in the tradition the Computer Assisted Mathematics Project (Johnson, Hatfield,

Katzman, Kieren, LaFrenz, & Walther, 1968). As recently as five years ago, I was a fervent supporter of having students study mathematics via computer programming. While I still have a great deal of empathy and respect for this approach, I now have reservations about programming as a vehicle for learning mathematics. The reservations come from the practicalities of having to teach a programming language and to teach programming constructs in order to teach mathematical concepts. If it is done well, it can work. However, I have strong suspicions that the requirements for making it happen are beyond the capabilities of the vast majority of teachers. To use programming as a vehicle for algorithmics requires more than knowledge of mathematics and knowledge of programming. It requires that that they be reflections of one another, and this is an intellectual achievement of the highest order.

I remain open to being shown how wrong I am. In fact, I would like to be wrong. I agree with Hatfield (chap. 11) on the potential benefits of student programming. However, there are many open questions as to what might constitute a proper balance among the components of content, curriculum, and programming, and there are many open questions about the long-term curricular and cognitive implications of student programming as a vehicle for learning mathematics.

There is another approach to introducing algorithmics into mathematical instruction. It also uses computers in non-traditional ways, but students do not write computer programs. Instead, they use programs that have been designed with the aim of making explicit the mathematical constraints of situations while imposing as few methodological constraints as possible. Students are asked to solve problems, posed independently of the program, but solve them within the constraints imposed *by* the program. A student's task is threefold: interiorize the mathematical constraints, develop methods for solving classes of problems, and construct a *recording scheme* that, at each step of his or her solution, reflects the state of the situation and a history of the steps taken so far. A student's recording scheme, viewed as an object in itself, is his or her personally-constructed algorithm for solving problems of a particular class.

However algorithmics is injected into mathematical instruction, one benefit seems clear. It is that students have occasions to experience the mutually-defining relationships between problems and methods. A slight modification in a "standard" problem may cause one to re-conceptualize one's methods for solving problems of that kind. Likewise, an analysis of a method may cause one to re-conceptualize one's understanding of the "kind" of problems for which the method was originally constructed. That is to say, a "class" of problems is determined by a method, but the boundaries of that class may not be as evident as originally thought. Likewise, problems give rise to methods, but methods may be more or less general than originally thought. Both realizations are propaedeutic for the construction of substantive mathematical knowledge.

Conceptions, Misconceptions, Teaching, and Learning

When reading Confrey's (chap. 8) paper, I was reminded of a recent event. Though my three-year-old daughter, Nicole, knows how to swim, when enrolled for summer swimming lessons, she could not get past "Station 1." Station 1 is where the component skills of swimming were taught. The reason she could not pass was this: instruction was focused so closely on component skills that Nicole never recognized them as part of her experience of swimming. Moreover, since instruction was focused so narrowly, the instructors never had an opportunity to see Nicole attempt component skills in the context of trying to swim. The instructors were teaching something of which Nicole was already capable, but Nicole never realized that she was already capable of what was being taught. As a result, instructors never had occasions to recognize areas where Nicole actually needed improvement. Gagné's reductionism has found a home in swimming instruction.

Confrey (chap. 8) gives us a holistic view of one students' attempts to make sense of the world of multiplicative structures, in general, and of exponential functions, in particular. She could have taken a reductionist, Gagné-like approach and interviewed Dan on "component" tasks, where her task analyses determined the component concepts. Had Confrey (chap. 8) done this, the result could easily have been analogous to Nicole's swimming lessons. We would have seen what Dan could do under the artificial constraints of an imposed segmentation of the concepts--a segmentation emanating from an analysis of instructional learning objectives. But we would not have known Dan's capabilities, nor would we have seen the simultaneous richness and poverty of Dan's cognitions in regard to multiplicative structures.

Confrey encouraged Dan to extend himself, to grope, to contradict himself, to make sense of nonsense and nonsense of sense. The picture we get of Dan's knowledge is not clean; it is quite messy. But in *our* making sense of Dan's struggles and inconsistencies, we learn more about the concept of exponential function. We see the tension between Dan's inclination to see the world additively (e.g., focusing on successive differences instead of on successive ratios) and the tension this causes when dealing with problems that, to him, were ostensibly multiplicative.

One task given to Dan ("Draw a picture of 5^3"), and along with Confrey's (chap. 8) subsequent analysis, raises an important question. To what extent is it profitable to portray exponentiation as repeated multiplication? If multiplication is portrayed as repeated addition, and if exponentiation is portrayed as repeated multiplication, then exponentiation reduces to repeated addition. If we are speaking about *calculating* some power of a whole number, the calculation does reduce to repeated addition. But calculation is not the issue. *Conception* is the issue. If one is attempting to conceive of an attribute of an object that we would normally call multiplicatively-structured and if that conception is fundamentally additively-structured, then one cannot conceive the attribute. We see this continually--children treating area as what we would normally call perimeter, or treating volume as what we would normally call surface area. We saw this in Dan's initial attempt to draw a picture of 2x3x4. We also saw his progressive "dimensionalization" of multiplication as he formed a product, and then created copies of that product. His pictures were still two-dimensional, but his comments indicated that he began imposing a structure on them that made his *conception* more like a Cartesian product.

Confrey's (chap. 8) analysis prompted me also to think about how 5^3 would fit with the portrayal of exponential functions as dimension-building objects. I do not know how people make sense of $5^{\sqrt{3}}$, if they make any sense of it at all.[4] Confrey has opened an intriguing and important domain of inquiry. I suspect it will be a long and fruitful research program.

Methodology

Kieren and Pirie (chap. 6) have taken the considerable task of developing a framework for viewing the evolution of a person's mathematical knowledge. As they say, "What is needed is an insightful way of viewing the whole of a person's growing mathematical knowledge and understanding built through this knowledge." They propose the use of recursive description as one way to capture the self-referentiality of knowing while remaining within the constraints of their view that humans are autopoeitic systems.

Their contribution is evidently ground-breaking, and as such requires careful reflection and analysis to realize its importance. It is a major contribution to the endeavor of providing useful and powerful visualizations of what we mean by "knowing mathematically." As with any work that breaks new ground, there are aspects to it that are problematic. This remark does not diminish the importance of Kieren and Pirie's (chap. 6) contribution. Rather, it frames my task of discussing their current work.

Explanations vs. Descriptions

There is a subtle yet significant difference between an explanation of an observation and a description of an observation. A description tells what happened; an explanation tells why what happened happened (and implicitly, why other things *didn't* happen). That is to say, an explanation must be framed by a theoretical context and oriented toward a class of possibilities, whereas a description is framed by an observation. Explanations and descriptions are evidently related, as descriptions are made out of the same theoretical stuff as explanations; but the intent of describing is different from the intent of explaining.

It is crucial to keep in mind the distinction between a description and an explanation as we struggle to understand students' mathematical thinking, for it is all too easy to unduly impute some of our constructs to our students. I am reminded of the recent literature on children's arithmetic, where early-on the counting strategy of "start with the larger" was described as children's growing awareness of the commutativity of addition (Ginsburg, Baroody, & Waxman, 1983). Lately this same strategy is being described in terms having nothing to do with the mathematical principle of commutativity (Baroody & Ginsburg, 1986). Did children begin to know less while behaving the same? No. Researchers began to do a better job of separating their mathematical knowledge from descriptions of children's mathematical knowledge.

The reason for dwelling on differences between descriptions and explanations is this: if our language of description is too powerful, or is used indiscriminately, our explanations may, in the end, be descriptions of our mathematics instead of children's mathematics (Steffe, 1988a).

Recursion vs. Repetition

Recursion and repetition are the primary constructs upon which procedural descriptions are founded. Dijkstra (1976) has given the classic distinction between repetition and recursion:

> The semantics of a repetitive construct can be defined in terms of a recurrence relation between *predicates*, whereas the semantic definition of general recursion requires a recurrence relation between *predicate transformers*. This shows quite clearly why I regard general recursion as an order of magnitude more complicated than just repetition. (Dijkstra, 1976, p. xvii, Italics in original)

Dijkstra used "predicate" to mean a proposition or a state of a computing mechanism and he used "predicate transformer" to mean a rule by which to derive one class of predicates from another. We can think of a predicate as a conception of a situation and a predicate transformer as a mental operation constituting a general relationship between classes of situations. In these terms, the distinction between repetition and recursion is between repeatedly applying an action to situations within a class (where attention is focused on the operation's "inputs") and the application of a mental operation to transform a class of situations (where attention is focused on the general relationship between the class of inputs and the class of outputs).

Unadvised use of recursive descriptions can result in more being attributed to students than we might wish. It is not uncommon to see recursion used where repetition would be more appropriate. I agree with Dijkstra when he says,

> Although correct [to define repetition in terms of recursion], it hurts me, for I don't like to crack an egg with a sledgehammer, no matter how effective the sledgehammer is for doing so. (Dijkstra, 1976, p. xvii)

The instance of Kieren and Pirie's (chap. 6) interpretation of Simon's and Alison's remarks in solving the handshake problem illustrates how the language of recursion can overpower observation. My understanding of *transcendence* is that it involves conceiving a process as if it were completed. However, by this criterion, neither Simon nor Alison transcended Joanne's original idea of having each of 35 people count handshakes and report their individual totals. Evidently, Simon constructed a solution method: compute the sum $34 + 33 + 32 + \ldots + 1$. Alison evidently generalized Simon's method to n people, the solution being $n + (n - 1) + \ldots + 1$. But neither appeared to transcend Joanne's original conception of each person walking down the line of people counting handshakes as they went. Appropriate attribution of transcendent recursion to a student's conception of the handshake problem would require evidence of thinking something like this:

> Each person in line leaves to drink a Pepsi, saying to the person on his or her left, "You and all the people on your left do all your handshakes. When you're done, tell me how many people were in your group and how many handshakes your group made. I'll get my answer by adding the number of people in your group to the number of handshakes."

It is important to realize that, were this method actually carried out, there would not be any actual handshakes--even though each person gives an explicit directive to do handshakes and count their number.

The difference between iterative and recursive conceptions of the handshake problem resides in how a solver thinks of a partial tally. In Simon's description, the person closest to the door computes the running tally, and only when the last person walks out is the tally identifiable with the total number of handshakes among some group of people. In the Pepsi-drinkers' method, each person in line computes a partial tally, and *any* partial tally is the total number of handshakes among some group of people. The distinction is significant.

Conventions for Attributions of Meaning

It is not clear what types of behavior can be taken as indicative of transcendent recursion. Watkins & Brazier (1985) reported that two students wrote procedures to produce derivatives of functions. Their procedures seemed to be highly recursive, yet it turned out that neither student had even the most rudimentary comprehension of recursion (Brazier, 1985).

How do we explain this apparent paradox? By appealing to Kieren's and Pirie's (chap. 6) conditions for transcendence. The students (using Logo) programmed a formula, $[f(g(x))]' = f'(g(x)) \cdot g'(x)$. The need for these students to transcend their initial conception of a derivative (as the output of the Logo function they were then defining) was short-circuited by their "knowledge" that the derivative already existed (as a *formula*) independently of the function they were defining. In short, they were not constructing a recursive function to make derivatives. They were translating a textbook formula into Logo.[5]

The larger issue is this: by what convention can we minimize the probability that we use too-powerful constructs to explain students' behaviors or to describe students' understandings? Steffe's group faced this problem in the course of developing what ultimately became known as counting types (Steffe, Thompson, & Richards, 1982; Steffe, von Glasersfeld, Richards, & Cobb, 1983). The methodological convention adopted was to attribute no more to a child's understandings than what is minimally necessary to account for their behavior on the task being performed, taking into consideration interpretations made of his or her performance on related tasks. Of course, to make this approach work, one must be conservative. To be conservative means to assume, at first, that a student does not fully understand the dimensions of the task. As an interpreter of behavior, one must always build a case for interpretations made.[6]

Cognitive Modeling

Steffe (chap. 3) and Dubinsky (chap. 9) have developed theoretical frameworks for modeling students' knowledge and describing its construction. However, the goals of their frameworks are quite different. The constructs in Steffe's (chap. 3) framework are used to explain children's construction of mathematical concepts within a specific domain at a high level of conceptual detail. In this volume, he used the notion of a unitizing operation to describe one student's construction of a number sequence.

Dubinsky's (chap. 9) framework, on the other hand, appears to have a different intent. It has the flavor of a "universal" framework for modeling the construction of many types of mathematical knowledge, from universal and existential quantification in the predicate calculus to mathematical induction.

Dubinsky's (chap. 9) approach is to apply his framework to his personal understandings of a concept. By doing this, he obtains a "genetic decomposition" of the concept, and takes that as his initial model for investigating students' concepts in the same domains. As Dubinsky (chap. 9) himself states, a genetic decomposition is only a starting point, to be discarded as one learns more about what students understand and how they came to these understandings. Even with this caveat in mind, however, I found myself asking for more than what this framework provided.

The framework, at least in the description given in Dubinsky's (chap. 9) paper, is too general to support precise interpretations of students' performance on tasks. I frequently found myself providing alternative interpretations to those given by Dubinsky, and the alternatives had more to do with the nature of students' understandings of specific concepts than with the broad operations of encapsulation, reversal, for example. Perhaps my understanding of the framework was not sufficiently precise to constrain my interpretations to those offered by Dubinsky (chap. 9). It might have helped were the central constructs of the framework somehow operationalized independently of their use in interpreting students' behavior. For example, what does it mean to "reverse a process"? Is it like running a motion picture backward? Is it like saying the alphabet backward? Or is it the performance of an operation that has the effect of undoing the effect of another? These are all very different from one another.

Mathematics Education as *Paideia*

The literary education portrayed by Lewin (chap. 10) aims to develop students' character by having them appropriate the literary artifacts of others. Through the dialectic of appropriation--viz., interiorization and internalization--students transform literary text into personally meaningful correlates of their own life experiences. In the process of appropriation, Lewin (chap. 10) says, students' personal schemas are likewise transformed.

> We have access not to [others] in the moments in which they create, but to their artifice, not to a life lived in its ongoing fullness, but to a stable presence, the artifact, generated out of that ongoingness. We ask students to engage these artifacts, enriched, educated in the literal sense of finding themselves drawn out. (p. 207)

How different is this from practices in mathematics education? On one hand, it is not different at all. Students are bombarded with artifacts in mathematics: definitions, theorems, proofs, and algorithms to name a few. On another hand, it is very different. The "objects" with which students are bombarded are not *presented* as artifacts. They are presented too often as "the truth," as being engraved in stone, having just arrived from the Mount.

What kind of mathematics pedagogy would parallel the pedagogy described by Lewin (chap. 10)? In teaching a definition, we would first need to make sure of its author.[7] Discussions would include why the author worded his or her definition this way and not another, why the boundaries were placed as they were.[8] Discussions of theorems would include why the conditions of the theorem are as given and not more, less, or differently restrictive. Discussions of algorithms would include the possible dependence of the algorithm's validity on the notational scheme in which it is expressed. In short, for mathematical pedagogy to parallel that described by Lewin, teachers would have to address the history of mathematics and the culture of the mathematical community explicitly. These issues traditionally are not an integral part of mathematics instruction or mathematics teacher preparation programs.

Lewin (chap. 10) gives us an extended example wherein he describes in concrete terms the process of appropriation as a dialectic of internalization and interiorization, and the subsequent transformations of self in relation to an objective Romanticism. As a reminder, here is the task he set for his students: (1) Write a passage explaining what it would mean for someone or something to be called "romantic." (2) Read "Mock on, Mock on, ..." . (3) Try to apply your description of "romantic" to see whether or not it fits the poem.

Lewin mentioned an essential ingredient of this task only in passing. *Students were told that Blake's poem is generally considered a good example of Romantic poetry.* Without their acceptance that Blake's poem "really is" romantic, students would only have had an occasion to classify it by their already-held criteria. They would not have found an occasion to reflect on what they understood by "romantic" in relation to meanings negotiated among the group of scholars who use the term precisely. That is, unless students accept characteristics of an artifact as being *objective*, they have little cause to constrain or reshape their understanding of it.

Students easily take notations as being objective (after all, they *see* them), but rarely do they consider that notations can have objective correlates.[9] That students eventually develop objective correlates of notation and notational transformations is a major aim of mathematics instruction. However, we are at a disadvantage. Students must relate experientially to anything before they can begin to objectify it, to apprehend it as immediately given. There are few mathematical artifacts that can be taken by students as objective independently of the notations in which they are expressed. One way to remedy this situation is to construct computer microworlds where things are made to happen by way of a notational scheme but whose behaviors are constrained according to the mathematical systems we wish students to construct. To interiorize the microworld is to interiorize the mathematical system by which the microworld's behavior is constrained (Thompson, 1985, 1987).

The major significance of Lewin's paper is in its implications for the preparation of mathematics teachers. Practitioners of mathematics primarily need to do mathematics. They do not need to relate to mathematical knowledge in the same way that a mathematics teacher needs to relate to it. Teachers must be able to express deeply principled knowledge "softly," so that its expression does not overpower a learner and yet is sensitive simultaneously to the learner's current constructions and to long-term curricular objectives. That is, a teacher of mathematics needs to be capable of imagining a construction of concepts that he or she already possesses and can be accessed only in their current moment. It is in this regard that I see the greatest relevance of Lewin's (chap. 10) notion of *paideia*, the formation of character.

How might we affect future teachers' reconstruction of the mathematics curriculum from a sequence of topics to a fabric of constructions? A slight paraphrase of one passage in Lewin's (chap. 10) paper addresses this question with remarkable acuity.

> [We] ask students to juxtapose their pre-existing ideas about [the curriculum] with their growing classroom familiarity of it in order to facilitate the examination, and perhaps revision, of their pre-suppositions. Unless a student can find the relevance of [the artifact], can construct its meaning for her or his own life [as a teacher], there is no possibility for *paideia*, for an engagement that enriches the self. (p. 212)

The key to affecting future teachers' reconstruction of the mathematics curriculum is in the creation of appropriate *artifacts* that somehow reflect a coherent vision of it as already recast into a fabric of constructions. A text (e.g., a book or an article) that elucidates an end-product of such a re-creation would be one possibility. A text, however, would suffer the same disadvantages as when the end-product of a mathematical construction in a notational system is expressed. Many teachers will have difficulty objectifying the ideas presented in the text; they will have difficulty constructing meaning for his or her own life as a teacher.

Another possibility is to extend the approach taken with computer microworlds. This is one that I am currently researching (Thompson, 1989a, 1989b). The general idea is to reflect the development of a conceptual field within a computer program, where "development" is reflected in transformations of constraints on the program's behavior. The program is designed so that it can be used to solve problems normally studied within the conceptual field, but the constraints on its behavior forces students to view the problems through the eyes of a student whose thinking is likewise constrained. The task ("Solve these problems") and the program are immediately taken by future teachers as objective. In the dialectic of internalizing and interiorizing the program, brought about through their use of it in solving problems, these future teachers have an occasion to reconstruct the conceptual field being studied.

Postscript

The papers in this volume clearly are concerned with mathematical experience. In many cases they describe "good" experiences to be had. However, we have not been informed about what *is* mathematical experience. As commentator, I could have taken "mathematical experience" as a theme to be drawn out from each paper. I chose not to try. The word "experience" was used to mean anything from a vague awareness to an exhilarating insight. I am not accusing the authors of using vague language. The problem is with common usage, which has strong realist overtones. "Experience" is sometimes used as if referring to an event (as separate from an experiencer), and at other times as if referring to the act of living through an event. Clearly, the latter was of concern to authors in this volume. To muddy the waters more, "experience" can be used as if referring to a totality of events lived through by someone, as in "Experience tells Paul that this will not work," and it can be used as if referring to the cumulative effect of living through those events, as in "Paul's experience is that this will not work." Perhaps "experience" is not a very useful word when trying to understand mathematical thinking through constructivism. Its non-technical usage overpowers any attempt at precision.

Notes

1. Preparation of this paper was supported in part by National Science Foundation grant number MDR 87-51381. Any opinions expressed are those of the author and do not necessarily reflect positions of the National Science Foundation.

2. Johnson-Laird (1988, pp. 134-137) notes, as does Bickhard, that one sign that Fodor's argument is wrong is that it proves that logic and concepts cannot even be innate because they cannot have evolved. However, Johnson-Laird takes a different approach than does Bickhard to refuting Fodor's argument generally and does so without explicitly denying encodingism. Instead, he draws a distinction

between an increase in a cognizing organizism's computational power (as modeled by a Turing machine) and an increase in its conceptual power. He argues that while the former cannot happen, the latter can.

3. The word "arbitrary" sometimes is given the same sense as "random." Clearly, this is not what is meant here. Instead, it means that there is nothing intrinsic to a symbol that would cause a loss of meaning by systematically substituting another one for it. We might find some symbols more practical or more convenient than others or easier to relate to other notations, but this is beside the point. For example, in about 20 minutes most adults can recast their number-name and numeration schemes using the notational and verbal scheme "a, b, c, d, e, f, g, h, i, ton, ton-a, ton-b, ton-c, ton-d, ton-e, ton-f, ton-g, ton-h, ton-i, b-ty, b-ty a, b-ty b, ...". That their original naming schemes were symbolic is suggested by the fact that they can perform their addition and subtraction algorithms using the new naming scheme. For children, number-names may or may not be symbols. They could be only indices of counting experiences or signs of having counted.

4. By "make sense" I do not mean to make formal sense, e.g. $5^{\sqrt{3}}$ as an equivalence class of Cauchy sequences. Rather, I wonder how they envision it as being the same sort of object as 5^3.

5. It is curious how different notations make it more or less difficult to translate the chain rule into Logo. Operator notation seems easiest to translate, i.e., $D_x (f o g) = [D_x(f) o g]\ D_x(g)$, but it is the most difficult for students to understand. Leibniz' notation, i.e., $dy/dx = dy/du \times du/dx$ seems easiest for students to understand but it is the hardest to translate into Logo. The prime notation used in this narrative seems somewhere in the middle on both counts.

6. On the other hand, interpreting will always be an art, and there are no guarantees against attributing too much or too little knowledge to another person's understanding.

7. Definitions do have authors, but their names are normally not mentioned in mathematics texts. The tradition in mathematics is that an author's name is attached only to important theorems or to important algorithms. Perhaps this is because of working mathematicians' implicit Platonism, that these things are "there," so only the hardest to find are named after the explorer finding them.

8. For example, why, in any definition of prime number, is 1 excluded from being prime? Because if it were included, the Fundamental Theorem of Arithmetic, which says that factorizations of integers are essentially unique, would have to be reworded in order to remain true, and the rewording would be very clumsy.

9. I deliberately use "notation" instead of "symbol." A notation is a mark of some kind; a symbol is something that has a referent. Students can see notations. If a student gives a notation meaning, for that student it is a symbol.

References

Alba, J. W., & Hasher, L. (1983). Is memory schematic? *Psychological Bulletin, 93,* 203-231.

Anderson, R. (1977). The notion of schemata and the educational enterprise. In R. Anderson, & W. Montague (Eds.), *Schooling and the acquisition of knowledge* (pp. 415-431). Hillsdale: Lawrence Erlbaum Associates.

Annas, J. (1986). Classical Greek Philosophy. In J. Boardman, J. Griffin, & O. Murray (Eds.), The *Oxford History of the Classical World* (pp. 204-233). Oxford: Oxford U. Press.

Antell, S. E., & Keating, D. P. (1983). Perception of numerical invariance in neonates. *Child Development, 54,* 695-701.

Barnes, B., & Bloor, D. (1986). Relativism, rationalism and the sociology of knowledge. In M. Hollis, & S. Lukes (Eds.), *Rationality and relativism* (pp. 21-47). Cambridge: MIT Press.

Baroody, A., & Ginsburg, H. (1986). The relationship between initial meaningful and mechanical performance in mathematics. In J. Hiebert (Ed.), *Conceptual and procedural knowledge: The case of mathematics* (pp. 75-112). Hillsdale, NJ: Erlbaum.

Bellah, R., Madsen, R., Sullivan, W., Swidler, A., & Tipton, S. (1985). *Habits of the heart.* Berkeley: University of California.

Bereiter, C. (1985). Toward a solution of the learning paradox. *Review of Educational Research, 55*(2), 201-226.

Berger, P. L., & Luckmann, T. (1967). *The social construction of reality.* New York: Doubleday & Company, Inc.

Berkeley, G. (1710). *A treatise concerning the principles of human knowledge.*

Bernstein, R. J. (1983). *Beyond objectivism and relativism: Science, hermeneutics, and praxis.* Philadelphia: University of Philadelphia Press.

Beth, E. W., & Piaget, J. (1966). *Mathematical epistemology and psychology.* Dordrecht, Reidel, (W. Mays, Trans. Original published 1965).

Bickhard, M. H. (1980a). *Cognition, Convention, and Communication.* New York: Praeger.

Bickhard, M. H. (1980b). A model of developmental and psychological processes. *Genetic Psychology Monographs, 102,* 61-116.

Bickhard, M. H. (1982). Automata theory, artificial intelligence, and genetic epistemology. *Revue Internationale de Philosophie, 36,* 549-566.

Bickhard, M. H. (1987). The social nature of the functional nature of language. In M. Hickmann (Ed.), *Social and Functional Approaches to Language and Thought.* Orlando: Academic.

Bickhard, M. H. (1988). Piaget on variation and selection models. *Human Development*, *31*, 274-312.

Bickhard, M. H., & Campbell, R. L. (1989). Interactivism and Genetic Epistemology. *Archives de Psychologie*, *57*, 99-121.

Bickhard, M. H., & Richie, D. M. (1983). *On the Nature of Representation*. New York: Praeger.

Bickhard, M. H., & Terveen, L. (1990). *The Impasse of Artificial Intelligence and Cognitive Science*. Unpublished manuscript.

Blevins, B. (1981). *The development of the ability to make transitive inferences*. Unpublished doctoral dissertation, University of Texas, Austin.

Bloom, B., Englehart, M., Furst, E., Hill, W., & Krathwohl, D. (1956). *A taxonomy of educational objectives: Handbook I, The cognitive domain*. New York: Longmans Green.

Bransford, J. D., & McCarrell, N. S. (1974). A sketch of a cognitive approach to comprehension: Some thoughts about understanding what it means to comprehend. In W. Weimer, & D. Palermo (Eds.), *Cognition and the symbolic processes* (pp. 189-229). Hillsdale, NJ: Erlbaum Associates, Inc.

Brazier, J. (1985). Private communication.

Breslow, L. (1981). Reevaluation of the literature on the development of transitive inferences. *Psychological Bulletin*, *89*, 325-351.

Bridgman, P. W. (1934). A physicist's second reaction to mengenlehre. *Scripta Mathematica*, 2, 101-117; 224-234.

Brousseau, G. (1984a). *Le rôle central du contrat didactique dans l'analyse et la construction des situation d'enseignment et d'apprentissage des mathématiques*. Ed. IMAG. Grenoble.

Brousseau, G. (1984b). Les obstacles épistémoloques et les problèmes en mathématiques. *Recherche en Didactiques des Mathématiques*, 4, (2), 164-198.

Brouwer, L. E. J. (1913). Intuitionism and formalism. *Bulletin of the American Mathematical Society*, *20*, 81-96.

Bruner, J. (1976). *Children's conception of addition and subtraction: The relation of formal and informal notions*. Unpublished doctoral dissertation, Cornell University.

Bruner, J. S. (1973). *The relevance of education*. New York: W.W. Norton & Company, Inc.

Bruner, J. S. (1987). Life as narrative. *Social Research*, *54*, 11-32.

Bruner, J. S. (1963). *The process of education.* New York: Vintage.

Brush, L. (1978). Preschool children's knowledge of addition and subtraction. *Journal for Research in Mathematics Education*, *9*, 44-54.

Bryant, P. E., & Trabasso, T. (1971). Transitive inferences and memory in young children. *Nature, 232*, 457-459.

Byers, B. (1984). Dilemmas in teaching and learning mathematics. *For the Learning of Mathematics*, 4(1), 35-39.

Campbell, R. L., & Bickhard, M. H. (1986). *Knowing levels and developmental stages*. Basel, Switzerland: Karger.

Campbell, R. L., Bickhard, M. H. (1987). A deconstruction of Fodor's anticonstructivism. *Human Development, 30*, 48-59.

Campbell, R., Cooper, R. G., & Blevins, B. (1983). *Development of extensions of addition/subtraction reasoning in elementary school children: Infinity, transfer, and connexity*. Paper presented at the Jean Piaget Society meeting, Philadelphia.

Caramuel, J. (1670). *Mathesis biceps, Meditatio prooemialis*. Campania: Officina Episcopali. (Italian translation by C. Oliva; Vigevano: Accademia Tiberina, 1977).

Case, R. (1978). Intellectual development from birth to adulthood: A neo-Piagetian interpretation. In R. S. Seigler (Ed.), *Children's thinking: What develops* (pp. 37-72)? Hillsdale, NJ: Erlbaum.

Case, R. (1985). *Intellectual development: Birth to adulthood*. London: Academic Press.

Chiari, G., & Nuzzo, M. L. (1988). Embodied minds over interacting bodies: A constructivist perspective on the mind-body problem. *International Journal of Psychology, 9*, 91-100.

Chomsky, N. (1980a). Discussion of Putnam's comments. In M. Piattelli-Palmarini (Ed.), *Language and learning: The debate between Jean Piaget and Noam Chomsky* (pp. 310-324). Cambridge: Harvard University Press.

Chomsky, N. (1980b). On cognitive structures and their development: A reply to Piaget. In M. Piattelli-Palmarini (Ed.), *Language and learning: The debate between Jean Piaget and Noam Chomsky* (pp. 35-52). Cambridge: Harvard University Press.

Cobb, P. (1987). Information-processing psychology and mathematics education: A constructivist perspective. *Journal of Mathematical Behavior, 6*, 3-40.

Cobb, P. & Steffe, L. (1983). The constructivist researcher as teacher and model builder. *Journal for Research in Mathematics Education, 14*, 83-94.

Cobb, P., Wood, T., & Yackel, E. (1991). A constructivist approach to second grade mathematics. In E. von Glasersfeld (Ed.), *Radical constructivism in mathematics education*. The Netherlands: Kluwer.

Cockcroft, W. (1988, April). *Mathematics and curriculum change in the U.K.*. Paper presented at the 1988 meeting of the National Council of Teachers of Mathematics, Chicago.

Cohler, B. (1982). Personal narrative and life course. In P. Baltes, & O. Brim (Eds.), *Life-span development and behavior* (pp. 205-241). Volume 4. New York: Academic.

Confrey, J. (1981). *Using the clinical interview to explore students' mathematical understandings*. A paper presented at the meeting of The American Educational Research Association, Los Angeles.

Confrey, J. (1985). Towards a framework for constructivist instruction. In L. Streefland (Ed.), *Proceedings of the Ninth International Conference for the Psychology of Mathematics Education* (pp. 477-483). Noordwijkerhout.

Confrey, J. (1986). *Misconceptions across the disciplines: Science, mathematics and programming*. A paper presented at the meeting of the American Educational Research Association, San Francisco.

Confrey, J. (1990). A review of the research on student conceptions in mathematics, science and programming. In C. Cazden (Ed.), *Review of Research in Education* (pp. 3-56), *16*, American Educational Research Association, Washington, DC.

Confrey, J. (1991). Learning to listen: A student 's understanding of powers of ten. In E. von Glasersfeld (Ed.), *Radical constructivism in mathematics education*. The Netherlands: Kluwer.

Cooper, R. G. (1984). Early number development: Discovering number space with addition and subtraction. In C. Sophian (Ed.), *Origins of cognitive skills* (pp. 157-192). Hillsdale, NJ: Erlbaum.

Cooper, R. G. (1985). The development of an understanding of relative numerosity in infants of 10 to 18 months. *Cahiers de Psychologie Cognitive*, *5*, 469.

Cooper, R. G. (1987a). *A longitudinal study of preschoolers' acquisition of transformation operators*. Paper presented at the International Society for the study of Behavioral Development, Tokyo, Japan, July.

Cooper, R. G. (1987b). *Infants' understanding of numerical concepts*. Paper presented at the Society of Research in Child Development, Baltimore, April.

Cooper, R. G., Campbell, R., & Blevins, B. (1983). Numerical representation from infancy to middle childhood: What develops? In D. R. Rogers, & J. A. Sloboda (Eds.), *The acquisition of symbolic skills* (pp. 519-533). New York: Plenum Press.

Csikszentmihalyi, M., & Beattie, O. (1979). Life themes: An exploration of their origins and effects. *J. Humanistic Psychology*, *19*, 45-63.

Culler, J. (1975). *Structuralist poetics*. Ithaca: Cornell.

Curtis, L., & Strauss, M. (1983). *Infant numerosity abilities: Discrimination and relative numerosity*. Paper presented at the Jean Piaget Society Meetings, Philadelphia.

Davis, P., & Hersh, R. (1981). *The mathematical experience*. Boston: Houghton Mifflin Company.

Davis, P., & R. Hersh (1986). *Descartes dream*. Boston: Harcourt Brace Jovanovich.

Davis, R. (1980). The postulation of certain, specific, explicit, commonly-shared frames. *Journal of Children's Mathematical Behavior, 3*, (1), 167-200.

Dawkins, R. (1987). *The blind watchmaker*. W.W. Norton, New York.

Derrida, J. (1982). Differance. In A. Bass (Trans.), *Margins of philosophy* (pp. 1-27). Chicago: University of Chicago.

Dewey, J. (1916). *Democracy and education*. New York: Macmillan.

Dijkstra, E. W. (1976). *A discipline of programming*. Englewood-Cliffs, NJ: Prentice-Hall.

DiSessa, A. (1983). Phenomenology and the evolution of intuition. In D. Gentner, & A. Stevens (Ed.), *Mental models* (pp. 15-33). Lawrence Erlbaum Press.

DiSessa, A. (1985). *Knowledge in pieces*. Address to the 15th annual symposium of the Jean Piaget Society. Philadelphia.

Douglas, R. (1985). One needs more than the algorithmic approach. *The College Mathematics Journal, 1*(16), 5-6.

Dreyfus, T. and Eisenberg, T. (1984). Intuitions on functions. *Journal of Experimental Education, 52*, (2), 77-85.

Dubinsky, E. (1986). Teaching mathematical induction I. *The Journal of Mathematical Behavior*, 5, 305-317.

Dubinsky, E. (1990). *Reflective abstraction in advanced mathematical thinking*. Unpublished manuscript.

Dubinsky, E. (in press *a*). On learning quantification. In M. S. Arora (Ed.), *Mathematics Education: The present state of the art*, UNESCO.

Dubinsky, E. (in press *b*). Teaching mathematical induction II. *The Journal of Mathematical Behavior.*

Dubinsky, E., Elterman, F. & Gong, C. (1988). The student's construction of quantification. *For the Learning of Mathematics, 8* (2), 44-51.

Dubinsky, E. & Lewin, P. (1986). Reflective abstraction and mathematics education: The genetic decomposition of inductin and compactness. *The Journal of Mathematical Behavior, 5*, 55-92.

Egan, K. (1987). Literacy and the oral foundations of education. *Harvard Educational Review, 57*, 445-472.

Eliot, T. (1920). Hamlet and his problems. In *The sacred wood* (pp. 95-103). London: Methuen.

Eves, H., & Newsom, C. V. (1958). *An introduction to the foundations and fundamental concepts of mathematics*. New York: Rinehart & Company, Inc.

Fabricius, W. V. (1983). Piaget's theory of knowledge: Its philosophical context. *Human Development*, *26*, 325-334.

Fann, K. T. (1970). *Peirce's theory of_abduction*. The Hague: Martinus Nijhoff.

Feng, C. (1973). *A course in algebra and trigonometry with computer programming*. Boulder, CO: The University of Colorado.

Fey, J. (Ed.). (1984). *Computing and mathematics: The impact on secondary school curricula*. Reston VA: National Council of Teachers of Mathematics.

Feyerabend, P. (1975). *Against method: Outline of an anarchistic theory of knowledge*. London: New Left Books.

Fish, S. (1980). *Is there a text in this class*? Cambridge: Harvard.

Flegg, G. (1983). *Numbers: their history and meaning*. Schocken Books, New York.

Fodor, J. (1975). *The language of thought*. New York: Crowell.

Fodor, J. (1980a). Discussion. In M. Piattelli-Palmarini (Ed.), *Language and learning: The debate between Jean Piaget and Noam Chomsky* (p. 269). Cambridge: Harvard University Press.

Fodor, J. (1980b). On the impossibility of acquiring "more powerful" structures. In M. Piattelli-Palmarini (Ed.), *Language and learning: The debate between Jean Piaget and Noam Chomsky* (pp. 142-162). Cambridge: Harvard University Press.

Fodor, J. (1981). The present status of the innateness controversy. In J. Fodor (Ed.), *Representations* (pp. 257-316). Cambridge: MIT Press.

Fodor, J. (1983). *The modularity of mind*. Cambridge: MIT Press.

Freeman, M. (1984). History, narrative, and life-span developmental knowledge. *Human Development*, *27*, 1-19.

Fuson, K., Stigler, J., & Bartsch, K. (1988). Grade placement of addition and subtraction topics in Japan, Mainland China, the Soviet Union, and the United States. *Journal for Research in Mathematics Education*, *19* (5), 449-456.

Gadamer, H. G. (1975). *Truth and method*. New York: Seabury.

Gardner, H. (1980). Cognition comes of age. In M. Piattelli-Palmarini, (Ed.) *Language and learning: The debate between Jean Piaget and Noam Chomsky* (pp. xix-xxxvi). Cambridge: Harvard University Press.

Geertz, C. (1983). *Local knowledge: Further essays in interpretive anthropology*. New York: Basic Books, Inc.

Gelman, R., & Baillargeon, L. (1983). A review of Piagetian concepts. In J. L. Flavell, & E. M. Markman (Volume Eds.) *Cognitive development: Handbook of Child Psychology* (pp. 167-230). New York: Wiley.

Gelman, R., & Gallistel, C. R. (1978). *The child's understanding of number*. Cambridge, MA: Harvard University Press.

Giere, R. N. (1988). *Explaining science: A cognitive approach*. Chicago: Chicago University Press.

Ginsburg, H., Baroody, A., & Waxman, B. (1983). Children's use of mathematical structure. *Journal for Research in Mathematics Education*, *14*, 156-168.

Glaser, R. (1984). Education and thinking. *American Psychologist*, *39*, 93-104.

Groen, G. J., & Resnick, L. B. (1977). Can preschool children invent addition algorithms? *Journal of Educational Psychology*, *69*, 645-652.

Halmos, P. (1985). Pure thought is better yet... *The College Mathematics Journal*, *1*(16), 14-16.

Hatfield, L. (1982). Instructional computing in mathematics teacher education. *Journal of Research and Development in Education*, *15* (4), 30-44.

Hatfield, L. (1985). Preparing teachers for instructional computing in school mathematics. In G. Jones (Ed.), *Proceedings of the Theme Group on Preservice Teacher Education, Fifth International Congress on Mathematics Education* (pp. 40-44). Brisbane, AUS: Kelvin Grove College of Advanced Education.

Hawkins, D. (1974a). I, Thou, and It. *The informed vision* (pp. 48-67). Agathon Press, New York.

Hawkins, D. (1974b). Nature, man and mathematics. *The informed vision* (pp. 109-131). Agathon Press, New York.

Hawkins, D., Apelman, M., Colton, R., & Flexner, A. (1982). *A report on research on critical barriers to the learning and understanding of elementary science*. NSF Contract #SED 80-08581.

Heidegger, M. (1962). *Being and time*. (J. Macquarrie, & E. Robinson, Trans.). New York: Harper & Row.

Hendrix, G. (1961). Learning by discovery. *Arithmetic Teacher*, *54*, 290-299.

Henry, N. B. (1942). *The psychology of learning: Forty-first yearbook of the NSSE*. Chicago: University of Chicago Press.

Hersh, R. (1986). Some proposals for revising the philosophy of mathematics. In T. Tymoczko (Ed.), *New Directions in the philosophy of mathematics* (pp. 9-28). Boston: Birkhauser.

Hilton, P. (1985). Algorithms are not enough... . *The College Mathematical Journal*, *1* (16), 8-9.

Hirsch, E.D. (1987). *Cultural literacy*. Boston: Houghton Mifflin.

Hockett, C.F. (1960). Logical considerations in the study of animal communication. In W. E. Lanzon, & W. N. Tavolga (Eds.), *Animal sounds and communication* (pp. 44-67). Washington, D.C.: American Institute of Biological Sciences.

Hogben, L. (1937). *Mathematics for the millions*. W. W. Norton, New York.

Inhelder, B., & Piaget, J. (1964). *Early growth of logic in the child*. New York: The Norton Library.

Iser, W. (1974). *The implied reader*. Baltimore: Johns Hopkins.

James, W. (1962). *Psychology (Briefer course)*. New York: Collier. (Originally published, 1892).

Johnson, D. K. (1989). *Toward a social constructivist epistemology* (Tech. Rep. No. 207). Amherst: University of Massachusetts, Scientific Reasoning Research Institute.

Kant, I. (1st edition, 1781). *Kritik der reinen Vernunft*. Berlin: Akademieausgabe, Vol. IV.

Kant, I. (1966). *The critique of pure reason*. (F. M. Muller, Trans.). Garden City, NY: Doubleday.

Kieren, T.E. (1988). Personal knowledge of rational numbers: Its intuitive and formal development. In J. Hiebert, & M. Behr (Eds.), *Number concepts in the middle grades* (pp. 62-80). Hillsdale: Lawrence Erlbaum Associates.

Kitching, G. (1988). *Marx and the philosophy of praxis*. New York: Routledge.

Klahr, D., & Wallace, J. G. (1976). *Cognitive development: An information-processing view*. Hillsdale: Lawrence Erlbaum Associates.

Krutetskii, V. (1976). *The psychology of mathematical abilities in schoolchildren*. University of Chicago Press, Chicago.

Kuhn, T. (1970). *The structure of scientific revolutions*. Chicago: The University of Chicago Press.

Lakatos, I. (1970). Falsification and the methodology of scientific research programmes. In I. Lakatos, & A. Musgrave (Eds.), *Criticism and the growth of knowledge* (pp. 91-196). Cambridge: Cambridge University Press.

Lakatos, I. (1976). *Proofs and refutations*. Cambridge: Cambridge University Press.

Laudan, L. (1977). *Progress and its problems*. Berkeley: U. of California Press.

Locke, J. (1690). *An essay concerning human understanding*.

MacIntyre, A. (1981). *After virtue*. Notre Dame: University of Notre Dame Press.

MacKay, D. (1969). *Information, meaning, and mechanism*. Cambridge, MA: MIT Press.

Margenau, H. (1987). *The miracle of existence*. Boston: New Sciences Library, Shambhala.

Markovitz, A., Eylon, B., & Bruckheimer, M. (1986). Functions today and yesterday. *For the Learning of Mathematics*, *6* (2), 18-24.

Mason, J., (with L. Burton, & K. Stacey). (1982). *Thinking mathematically*. London: Addison-Wesley Publishing Company.

Maturana, H. (1978). Biology of language: The epistemology of reality. In G. A. Miller, & E. Lenneberg (Eds.), *Psychology and biology of language and thought* (pp. 27-63). New York: Academic Press.

Maturana, H. (1988). *Ontology of observing: The biological foundations of self consciousness and the physical domain of existence.* Paper presented at the conference of the American Society for Cybernetics, "Texts in Cybernetic Theory," Felton, CA.

Maturana, H. R., & Varela, F. J. (1980). *Autopoiesis and cognition.* Dordrecht, Holland: Reidel.

Maturana, H. R., & Varela, F. J. (1987). *The tree of knowledge.* Boston; New Science Library.

Maurer, S. (1983). The effects of a new college mathematics curriculum on high school mathematics. In A. Ralston, & G. Young (Eds), *The future of college mathematics* (pp. 153-76). New York: Springer-Verlag.

Maurer, S. (1984). Two meanings of algorithmic mathematics. *Mathematics Teacher,* 6(77), 430-435.

Maurer, S. (1985). The algorithmic way of life is best. *The College Mathematics Journal, 1*(16), 2-4, 17-18.

McDermott, L. (1984). Research on conceptual understanding in mechanics. *Physics Today, 37,* 24-32.

McLellan, J.A., & Dewey, J. (1895). *The psychology of number.* New York: Appleton.

Menninger, K. (1969). *Number words and number symbols: A cultural history of numbers.* Cambridge: The M. I. T. Press.

Mischler, E. (1986). *Research interviewing: Context and narrative.* Harvard University Press, Cambridge.

Moessinger, P., & Poulin-Dubois, D. (1981). Piaget on abstraction. *Human Development, 24,* 347-353.

Moise, E. E. (1963). *Elementary geometry from an advanced standpoint.* Reading: Addison-Wesley Publishing Company, Inc.

Monk, S. (1987). *Student's understanding of functions in calculus.* Unpublished manuscript. University of Washington, Seattle.

Morley, A. (1981). Teaching and learning algorithms. *For the Learning of Mathematics, 2*(2), 50-51.

Neisser, U. (1976). *Cognition and reality.* San Francisco: W. H. Freeman and Company.

Newman, J. (1956). *The world of mathematics, Vol. 3* (pp. 1623-1626). Simon and Schuster, New York.

Nisbett, R., & Ross, L. (1985). Judgmental heuristics and knowledge structures. In H. Kornblith (Ed.), *Naturalizing epistemology* (pp. 189-215). Cambridge: MIT Press.

Norman, D. (1980). What goes on in the mind of the learner. In W. McKeachie (Ed.), New directions for teaching and learning. *Learning, Cognition, and College Teaching* (pp. 37-49), *2*, San Francisco: Jossey-Bass.

Olson, D. (1980). Some social aspects of meaning in oral and written language. In D. Olson (Ed.), *The social foundations of language and thought* (pp. 90-108). New York: Norton.

Opper, S. (1977). Piaget's clinical method. *The Journal of Children's Mathematical Behavior, 1*, (4), 90-107.

Osborne, A. (1974). Conditions for algorithmic imagination. In M. Suydam & A. Osborne (Eds.), *Algorithmic learning* (pp. 13-22). Columbus OH: ERIC/SMEAC.

Papert, S. (1980). *Mindstorms: Children, computers, powerful ideas*. New York: Basic Books, Inc.

Piaget, J. (1929). Les deux direction de la pensée scientifique, *Archives des Sciences Physiques et naturelles, 11*, 145-165.

Piaget, J. (1937). *La construction du réel chez l'enfant'*. Neuchâtel: Delachaux et Niestlé.

Piaget, J. (1945). *La formation du symbole chez l'enfant*. Paris: Delachaux et Niestlé.

Piaget, J. (1950). *The psychology of intelligence.* London: Routledge & Kegan-Paul.

Piaget, J. (1962). *Play, dreams, and imitation in childhood*. (C. Gattegno & F. Hodgson, Trans.) New York: Norton.

Piaget, J. (1967). Le système et la classification des sciences. In J. Piaget (Ed.), *Logique et Connaissance scientifique*, (1151-1224). Paris: Encyclopédie de la Pléiade, Gallimard.

Piaget, J. (1968). *Six psychological studies.* New York: Vantage Books.

Piaget, J. (1969). *The mechanisms of perception*. (G. Seagrim, Trans.) New York: Basic Books.

Piaget, J. (1970a). *Genetic epistemology*. New York: Columbia University Press.

Piaget, J. (1970b). *Tendences principales de la recherche dans les sciences sociales et humaines*. Paris/The Hague: Mouton.

Piaget, J. (1971). *Biology and Knowledge*. (B. Walsh, Trans.) Chicago: University of Chicago Press. (Original published 1967).

Piaget, J. (1973). *To understand is to invent: The future of education*. New York: Grossman.

Piaget, J. (1974a). *Adaptation vitale et psychologie de l'intelligence*. Paris: Hermann.

Piaget, J. (1974b). *La prise de conscience*. Paris: Presses Universitaires de France.

Piaget, J. (1974c). *Réussir et comprendre*. Paris: Presses Universitaires de France.

Piaget, J. (1974/1980). *Experiments in contradiction*. Chicago: University of Chicago Press.

Piaget, J. (1975). *L'équilibration des structures cognitives*. Paris: Presses Universitaires de France.

Piaget, J. (1980a). *Adaptation and Intelligence*. (S. Eames, Trans) Chicago: University of Chicago Press. (Original published 1974).

Piaget, J. (1980b). The psychogenesis of knowledge and its epistemological significance. In M. Piattelli-Palmarini (Ed.), *Language and learning: The debate between Jean Piaget and Noam Chomsky* (pp. 23-34). Cambridge: Harvard University Press.

Piaget, J. (1985). *The equilibration of cognitive structures*. (T. Brown & K. Thampy, Trans.) Chicago: University of Chicago Press.

Piaget, J., & collaborators, (1977). *Recherches sur l'abstraction réfléchissante*, Vol. 1 & II. Paris: Presses Universitaires de France.

Piaget, J. & Garcia, R. (1983). *Psychogenèse et histoire des sciences*. Paris. Flammarion.

Piaget, J., Grize, J. B., Szeminska, A., & Vinh-Bang, A. (1977). *Epistemology and psychology of functions*. Dordrecht: D. Reidel.

Piattelli-Palmarini, M. (1980). *Language and learning: The debate between Jean Piaget and Noam Chomsky*. Cambridge: Harvard University Press.

Pirie, S.E.B. (1987). *Mathematical investigations in your classroom*. Basingstoke: Macmillan Educational Press.

Pirie, S.E.B. (1988) Understanding: Instrumental, relational, intuitive, constructive, formalized... How can we know? *For the Learning of Mathematics*, *8*, 3,2-6.

Plunkett, S. (1981). Fundamental questions for teachers. *For the Learning of Mathematics*, *1*(3), pp. 46-48.

Polya, G. (1957). *How to solve it* (2nd ed). Garden City, NY: Doubleday.

Popper, K. (1959). *The logic of scientific discovery*. New York: Harper.

Popper, K. (1965). *Conjectures and refutations*. New York: Harper.

Popper, K. (1972). *Objective knowledge*. London: Oxford Press.

Popper, K. R. (1982). *Quantum theory and the schism in physics*. Totowa, NJ: Rowman and Littlefield.

Popper, K. (1985). The problem of induction. In D. Miller (Ed.), *Popper Selections* (pp. 101-117). Princeton: Princeton U. Press.

Porter, A. (1989). A curriculum out of balance: The case of elementary mathematics. *Educational Researcher*, *18*(5), 9-15.

Powers, W. (1973). *Behavior: The control of perception*. Chicago: Aldine.

Powers, W. (1978). Quantitative analysis of purposive systems: Some spadework at the foundations of scientific psychology. *Psychological Review*, *85*, 417-435.

Renz, P. (1985). The path to hell. *The College Mathematical Journal*, 1 (16), pp. 9-11.

Ricoeur, P. (1967). *The symbolism of evil*. (E. Buchanan, Trans.) Boston: Beacon Press.

Ricoeur, P. (1976). *Interpretation theory: Discourse and the surplus of meaning*. Fort Worth, Texas: Texas Christian Univ.

Riley, C. A., & Trabasso, T. (1974). Comparatives, logical structures, and encoding in a transitive inference task. *Journal of Experimental Child Psychology*, *45*, 972-977.

Riley, M. S., Greeno, J. G., & Heller, J. L. (1983). Development of children's problems solving abillity in arithmetic. In H. P. Ginsburg (Ed.), *The development of mathematical thinking* (pp. 153-196). New York: Academic Press.

Rotenstreich, N. (1974). Humboldt's prolegomena to philosophy of language. *Cultural Hermeneutics*, *2*, 211-227.

Rozin, P. (1976). The evolution of intelligence and access to the cognitive unconscious. In J. M. Sprague, & A. N. Epstein (Eds.), *Progress in psychobiology and physiological psychology* (pp. 1-16) (Vol. 6). New York: Academic Press.

Segal, L. (1986). *The dream of reality: Heinz von Foerster's constructivism*. New York: Norton.

Shatz, M. (1978). The relationship between cognitive processes and the development of communication skills. In C. B. Keasey (Ed.), *Nebraska symposium on motivation* (pp. 1-42) (Vol. 26). Lincoln: University of Nebraska Press.

Shulman, L. (1978). *Critical research sites*. An invited address to the SIG/Research in Mathematics Education at the meeting of the American Educational Research Association, Toronto.

Smith, L. (1981). Piaget mistranslated, *Bulletin of the British Psychological Society*, *4*, 1-3.

Starkey, P., & Cooper, R. (1980). Perception of numbers by human infants. *Science*, *210*, 1033-1035.

Starkey, P., Spelke, E., & Gelman, R. (1983). Detection of one-to-one correspondence by human infants. *Science*, *222*, 179-181.

Steffe, L. P. (1988a). Children's construction of number sequences and multiplying schemes. In M. Behr, & J. Hiebert (Eds.), *Number concepts and operations in the middle grades* (pp. 119-140). Hillsdale, NJ: Lawrence Erlbaum, 119-140.

Steffe, L. P. (1988b). Lexical and syntactical meanings of Brenda, Tarus, and James. In L. P. Steffe, & P. Cobb (with E. von Glasersfeld), *Construction of arithmetical meanings and strategies* (pp. 95-147). New York: Springer-Verlag.

Steffe, L. P. (1988c). Modifications of the counting scheme. In L. P. Steffe, & P. Cobb (with E. von Glasersfeld), *Construction of arithmetical meanings and strategies* (pp. 284-322). New York: Springer-Verlag.

Steffe, L. P. (1991). The constructivist teaching experiment. In E. von Glasersfeld (Ed.), *Radical constructivism in mathematics education*. The Netherlands: Kluwer.

Steffe, L. P., & Cobb, P. (with E. von Glasersfeld). (1988). *Construction of arithmetical meanings and strategies*. New York: Springer-Verlag.

Steffe, L. P., Hirstein, J., & Spikes, W. C. (1975). *Quantitative comparisons and class inclusion as readiness variables for learning first-grade arithmetical content*. (Technical Report No. 9). Tallahassee, Florida: Project for Mathematical Development of Children, 1976. (ERIC Document Reproduction Service No. Ed 144 808).

Steffe, L. P., Thompson, P., & Richards, J. (1982). Children's counting in arithmetical problem solving. In T. Carpenter, T. Romberg, & J. Moser (Eds.), *Children's arithmetic: A cognitive perspective* (pp. 83-98). Hillsdale, NJ: Lawrence Erlbaum.

Steffe, L. P., & von Glasersfeld, E. (1988). On the construction of the counting scheme. In L. P. Steffe, & P. Cobb (with E. von Glasersfeld), *Construction of arithmetical meanings and strategies* (pp. 1-19), New York: Springer-Verlag.

Steffe, L. P., von Glasersfeld, E., Richards, J., & Cobb, P. (1983). *Children's counting types: Philosophy, theory, and application*. New York: Praeger.

Stigler, J., Fuson, K., Ham, M., & Kim, M. (1986). An analysis of addition and subtraction word problems in American and Soviet elementary mathematics textbooks. *Cognition and Instruction*, *3*(3), 153-171.

Strauss, M., & Curtis, L. (1981). Infant perception of numerosity. *Child Development*, *52*, 1146-1152.

Strauss, M., & Curtis, L. (1984). Development of numerical concepts in infancy. In C. Sophian (Ed.), *Origins of cognitive skills* (pp. 131-155). Hillsdale, NJ: Erlbaum.

Suchting, W. A. (1986). *Marx and philosophy*. London: MacMillan Press.

Suppes, P., & Groen, G. (1966). *Some counting models for first-grade performance data on simple addition facts*. Institute for Mathematical Studies in the Social Sciences. Stanford: Stanford University.

Suydam, M. (1974). Algorithmic learning: Introduction. In M. Suydam, & A. Osborne (Eds.), *Algorithmic learning* (pp. 3-7). Columbus OH: ERIC/SMEAC.

Thompson, P. (1985). Experience, problem solving, and learning mathematics: Considerations in developing mathematics curricula. In E. Silver (Ed.), *Teaching and learning mathematical problem solving: Multiple research perspectives* (pp. 417-436). Hillsdale: NJ Lawrence Erlbaum Associates.

Thompson, P. (1987). Mathematical microworlds and intelligent computer-assisted instruction. In G. Kearsley, (Ed.), *Artificial intelligence and instruction* (pp. 83-104). Addison-Wesley.

Thompson, P. (1989a). *A cognitive model of quantity-based algebraic reasoning.* Paper presented at the Annual Meeting of the American Educational Research Association, San Francisco, March 22-27.

Thompson, P. W. (1989b) Artificial intelligence, advanced technology, and learning and teaching algebra. In C. Kieren & S. Wagner (Eds.) *Research issues in the learning and teaching of algebra* (pp. 135-161). Reston, VA: National Council of Teachers of Mathematics. [Pre-printed in *Learning and Technology*, 1988, 2(5&6).]

Trabasso, T. (1975). Representation, memory, and reasoning: How do we make transitive inferences? In A. D. Pick (Ed.), *Minnesota symposia on child psychology* (pp. 72-168) (Vol. 9). Minneapolis: University of Minnesota Press.

Vergnaud, G. (1983). Multiplicative structures. In R. Lesh, & M. Landau (Eds.), *Acquisition of mathematical concepts and processes* (pp. 127-174). New York: Academic Press.

Vinner, S. (1983). Concept definition, concept image and the notion of function. *International Journal for Mathematics Education, Science and Technology, 14*, (3), 293-305.

Vitale, B. (1989). Elusive recursion: A trip in a recursive land. *New Ideas in Psychology*, 7 (3), 253-276.

von Foerster, H. (1965). Memory without record. In D.P. Kimble (Ed.), *The anatomy of memory* (pp. 388-432), Palo Alto, CA.: Science and Behavior Books. (Reprinted in H. von Foerster, *Observing systems*, Salinas, CA.: Intersystems Publications, 1981.)

von Foerster, H. (1984). On constructing reality. In P. Watzlawick (Ed.), *The invented reality* (pp. 41-61). New York: Norton.

von Glasersfeld, E. (1974). Signs, communication, and language. *Journal of Human Evolution, 3*, 465-474. (Reprinted in von Glasersfeld, E. (1987). *The construction of knowledge*. Salinas, CA.: Intersystems Publications.)

von Glasersfeld, E. (1976). Cybernetics and cognitive development. *American Society of Cybernetics Forum, 8* (3&4), 115-120.

von Glasersfeld, E. (1977). Linguistic communication: Theory and definition. In D. Rumbaugh et al. (Eds.), *Language learning by a chimpanzee: The Lana project* (pp. 55-71). New York: Academic Press.

von Glasersfeld, E. (1978). Cybernetics, experience, and the concept of self. In M. N. Ozer (Ed.), *Toward the more human use of human beings.* (pp. 109-122). Boulder, CO: Westview Press.

von Glasersfeld, E. (1981). An attentional model for the conceptual construction of units and number, *Journal for Research in Mathematics Education, 12*(2), 83-94.

von Glasersfeld, E. (1982). An interpretation of Piaget's constructivism. *Revue Internationale de Philosophie, 36*(4), 612-635.

von Glasersfeld, E. (1983). Learning as constructive activity. In J. C. Bergeron & N. Herscovics (Eds.), *Proceedings of the Fifth Annual Meeting of the North American Chapter of the International Group for the Psychology of Mathematics Education* (pp. 41-68). PME-NA: Montreal, Canada.

von Glasersfeld, E. (1984). An introduction to radical constructivism. In P. Watzlawick (Ed.), *The invented reality* (pp. 17-40). New York: Norton.

von Glasersfeld, E. (1987). Preliminaries to any theory of representation. In C. Janvier (Ed.), *Problems of representation in the teaching and learning of mathematics* (pp. 215-225). Hillsdale, N.J.: Lawrence Erlbaum.

von Glasersfeld, E., Steffe, L. P., & Richards, J. (1983). An analysis of counting and what is counted (pp. 21-44). In L. P. Steffe, E. von Glasersfeld, J. Richards, & P. Cobb. *Children's counting types: Philosophy, theory, and application*. New York: Praeger.

von Humboldt, W. (1907). *Werke*, (Vol. 7, part 2). Berlin: Leitmann.

Vuyk, R. (1981). *Piaget's genetic_epistemology 1965-1980* (Vol. I & II). New York: Academic Press.

Wales, B. (1984). *A study of children's language use when solving partitioning problems: Grades two through four.* Unpublished masters thesis, University of Alberta, Edmonton.

Watkins, W., & Brazier, J. (1985). The chain rule in the Logo environment. *Proceedings of the Psychology of Mathematics Education-North America, 1*, 363-367.

Winnicott, D. (1971). Transitional objects and transitional phenomena. In *Playing and reality* (pp. 1-30). Harmondsworth: Penguin.

Wittrock, M. (1973). *Recent research in cognition applied to mathematics learning*. Columbus OH: ERIC/SMEAC.

Author Index

Subject Index